MODELING WITH AN ANALOG HARDWARE DESCRIPTION LANGUAGE

THE KLUWER INTERNATIONAL SERIES IN ENGINEERING AND COMPUTER SCIENCE

ANALOG CIRCUITS AND SIGNAL PROCESSING
Consulting Editor
Mohammed Ismail
Ohio State University

Related Titles:

LOW-VOLTAGE CMOS OPERATIONAL AMPLIFIERS: *Theory, Design and Implementation*, Satoshi Sakurai, Mohammed Ismail
 ISBN: 0-7923-9507-7
ANALYSIS AND SYNTHESIS OF MOS TRANSLINEAR CIRCUITS, Remco J. Wiegerink
 ISBN: 0-7923-9390-2
COMPUTER-AIDED DESIGN OF ANALOG CIRCUITS AND SYSTEMS, L. Richard Carley, Ronald S. Gyurcsik
 ISBN: 0-7923-9351-1
HIGH-PERFORMANCE CMOS CONTINUOUS-TIME FILTERS, José Silva-Martínez, Michiel Steyaert, Willy Sansen
 ISBN: 0-7923-9339-2
SYMBOLIC ANALYSIS OF ANALOG CIRCUITS: Techniques and Applications, Lawrence P. Huelsman, Georges G. E. Gielen
 ISBN: 0-7923-9324-4
DESIGN OF LOW-VOLTAGE BIPOLAR OPERATIONAL AMPLIFIERS, M. Jeroen Fonderie, Johan H. Huijsing
 ISBN: 0-7923-9317-1
STATISTICAL MODELING FOR COMPUTER-AIDED DESIGN OF MOS VLSI CIRCUITS, Christopher Michael, Mohammed Ismail
 ISBN: 0-7923-9299-X
SELECTIVE LINEAR-PHASE SWITCHED-CAPACITOR AND DIGITAL FILTERS, Hussein Baher
 ISBN: 0-7923-9298-1
ANALOG CMOS FILTERS FOR VERY HIGH FREQUENCIES, Bram Nauta
 ISBN: 0-7923-9272-8
ANALOG VLSI NEURAL NETWORKS, Yoshiyasu Takefuji
 ISBN: 0-7923-9273-6
ANALOG VLSI IMPLEMENTATION OF NEURAL NETWORKS, Carver A. Mead, Mohammed Ismail
 ISBN: 0-7923-9049-7
AN INTRODUCTION TO ANALOG VLSI DESIGN AUTOMATION, Mohammed Ismail, José Franca
 ISBN: 0-7923-9071-7
INTRODUCTION TO THE DESIGN OF TRANSCONDUCTOR-CAPACITOR FILTERS, Jaime Kardontchik
 ISBN: 0-7923-9195-0
VLSI DESIGN OF NEURAL NETWORKS, Ulrich Ramacher, Ulrich Ruckert
 ISBN: 0-7923-9127-6
LOW-NOISE WIDE-BAND AMPLIFIERS IN BIPOLAR AND CMOS TECHNOLOGIES, Z. Y. Chang, Willy Sansen
 ISBN: 0-7923-9096-2
ANALOG INTEGRATED CIRCUITS FOR COMMUNICATIONS: Principles, Simulation and Design, Donald O. Pederson, Kartikeya Mayaram
 ISBN: 0-7923-9089-X
SYMBOLIC ANALYSIS FOR AUTOMATED DESIGN OF ANALOG INTEGRATED CIRCUITS, Georges Gielen, Willy Sansen
 ISBN: 0-7923-9161-6

MODELING WITH AN ANALOG HARDWARE DESCRIPTION LANGUAGE

by

H. Alan Mantooth
Mike Fiegenbaum
Analogy, Inc.

KLUWER ACADEMIC PUBLISHERS
Boston / Dordrecht / London

Distributors for North America:
Kluwer Academic Publishers
101 Philip Drive
Assinippi Park
Norwell, Massachusetts 02061 USA

Distributors for all other countries:
Kluwer Academic Publishers Group
Distribution Centre
Post Office Box 322
3300 AH Dordrecht, THE NETHERLANDS

Library of Congress Cataloging-in-Publication Data
A C.I.P. Catalogue record for this book is available
from the Library of Congress.

Copyright © 1995 by Analogy, Inc.

All rights reserved. No part of this publication may be reproduced, stored in a retrieval system, or transmitted, in any form or by any means, electronic, mechanical, photocopying, recording, or otherwise, without the prior written permission of the publisher, Kluwer Academic Publishers, 101 Philip Drive, Assinippi Park, Norwell, Massachusetts 02061 USA. Printed in the United States of America.

SPICE is a circuit simulation program developed at the University of California, Berkeley
HDL-A ™ is a trademark of Anacad, Inc.
SpectreHDL ™ is a trademark of Cadence Design Systems, Inc.

Analogy, Inc. claims the following patents and trademarks on the Saber simulator, the MAST modeling language, and their associated products:

U.S. Patent Nos. 4,985,860 (Canadian Patent No. 1,319,989), 5,092,780.

Analogy® and MAST® are registered trademarks of Analogy, Inc. AHDL® and Analogy HDL® are registered trademarks of Analogy, Inc. in the U.K. and Germany.

SABER® is a registered trademark of American Airlines, Inc., licensed to Analogy, Inc.

LIMITS OF LIABILITY AND DISCLAIMER OF WARRANTY
The authors and publisher have exercised care in preparing this book and the programs contained in it. They make no representation, however, that the programs are error-free or suitable for every application to which the reader may attempt to apply them. The authors and publisher make no warranty of any kind, expressed or implied, with regard to these programs or the documentation or theory contained in this book, all of which are provided "as is." The authors and publisher shall not be liable for damage in connection with, or arising out of the furnishing, performance, or use of these programs or the associated descriptions or discussions.

Readers should test any program in their own systems and compare the results with those presented in this book. They should then construct their own test programs to verify that they fully understand the requisite conventions and formats for each program. They should then test the specific application thoroughly.

To Opie and DeLois Mantooth
and the memory of Marie Mantooth

Alan Mantooth

To the memory of Bill and Alice Fiegenbaum

Mike Fiegenbaum

Contents

Preface		xv
Section 1	***Fundamentals of Modeling***	*1*
1.	**Modeling: A Working Definition**	**3**
	1.1 What Is Simulation?	3
	Simulation and modeling: the user's perspective	7
	1.2 What Is a "Model" Anyway?	8
	A MOSFET is a MOSFET is a MOSFET...or is it?	10
	1.3 References	12
2.	**Modeling With an Analog HDL**	**13**
	2.1 What Is an AHDL?	13
	2.2 Modeling Concepts	14
	Mathematics	14
	Conceptualizing a model	16
	2.3 What Can Really be done with an Analog HDL?	18
	Other technologies	18
	Modeling hierarchy	18
	Macromodeling	20
	Behavioral modeling	20
	Semiconductor device modeling	21
	Model validation	23
	Standardization	25
	2.4 References	26

3.	**Macromodeling**		27
	3.1	Features of Macromodeling	27
		Circuit simplification	28
		Circuit build-up	28
		Symbolic macromodeling	29
		Combining techniques	29
	3.2	Example—Simultaneous Differential Equations	32
		SPICE implementation	32
		Behavioral implementation in MAST	34
		Conclusion	35
	3.3	References	35
4.	**Top-Down Design**		37
	4.1	Introduction	37
	4.2	An Electro-mechanical Motion Control System	39
		Simulating the top-level design	40
		Moving down—refining the design	41
		Investigating a solution for backlash	43
		Full system implementation	45
		Conclusions	47
	4.3	References	47

Section 2	*Model Implementation*		49
5.	**Connecting Models to the MAST AHDL**		51
	5.1	What is a Template?	51
		A few definitions	52
		A quick example	53
	5.2	Template Organization and the Modeling Process	53
		Basic template organization	53
		Dependent and independent variables	54
		Philosophy of writing a simulation model	55
		Expressing the basic model	56
	5.3	Template Sections	59
		Comments	59

		Header	59
		Header declarations	61
		Template body	61
		Local declarations	61
		Parameters section	62
		Netlist section	62
		When statements	63
		Values section	63
		Control section	63
		Equations section	64
		Simplified MAST—sections can be optional	64
6.	**Netlists—Using Hierarchy**		**67**
	6.1	Modularity and Hierarchy	67
	6.2	Constructing a Module for an Active Filter	70
		Netlist usage	70
		Creating a hierarchical template	72
		Conclusion	75
7.	**Basic Linear Devices**		**77**
	7.1	Constant Current Source	77
	7.2	Resistor	79
		If constructs (condition checking)	80
	7.3	Capacitor—Using Differentiation	82
		Intermediate variables	82
		Differentiation	83
	7.4	Constant Voltage Source	84
	7.5	Inductor—Using Integration	85
8.	**Piecewise Linear and Table Lookup Modeling**		**87**
	8.1	Nonlinear Modeling	88
	8.2	Modeling a Simple Voltage Limiter	89
		Characteristic equations	89
		Template header and header declarations	91
		Local declarations	91
		Template equations	91

		Control section	93
	8.3	Modeling a Nonlinear Conductance	94
		Conclusions	96
	8.4	References	96
9.		**Modeling Nonlinear Devices**	**97**
	9.1	Modeling a p-n Junction Diode	98
		Diode equations	98
		The large-signal topology	99
		Diode parameters	100
		Diode template	101
		Header and header declarations	103
		Local declarations	103
		Parameters section	104
		Values section	104
		Equations section	104
		Control section	105
		Simplified template form	105
	9.2	Modeling a MOSFET	106
		MOSFET equations	106
		The large-signal topology	108
		MOSFET parameters	109
		MOSFET template	110
		Header and header declarations	115
		Local declarations	116
		Parameters section	116
		Values section	117
		Equations section	117
		Control section	117
		Operational characteristics	118
		Conclusions	118
	9.3	References	119
10.		**Digital Modeling**	**121**
	10.1	Digital Characteristics	121
	10.2	Digital Inverter Template	123

		Digital connection points	124
		Digital values	124
		The when statement	125
		The schedule_event statement	126
	10.3	Initialization	127
	10.4	Digital Clock Template	127
		Header and Local state variables	129
		Initialization	129
	10.5	Conflict Resolution and the Driven Function	130
		Implementing conflict resolution	132
		Conclusions	132
11.	**Modeling Mixed Analog-Digital Systems**		**133**
	11.1	Modeling a Voltage Comparator	134
	11.2	A Digitally-Controlled, Ideal Switch	140
		Analog time steps	142
		Example circuit	143
		Conclusions	144
	11.3	References	144
12.	**Advanced AHDL Capabilities**		**145**
	12.1	Control System Modeling (s-Domain)	146
		Creating basic control system templates	148
		Laplace domain modeling	150
	12.2	Sampled Data System Modeling (z-Domain)	157
	12.3	Statistical Modeling	159
		Probability density function (PDF)	160
		Cumulative density function (CDF)	163
		Correlation	163
		The random function	165
	12.4	Noise Modeling	165
		Adding noise to the diode template	168
		Conclusions	170
	12.5	References	170

| Section 3 | Advanced Applications | 171 |

13. Electro-mechanical Systems — 173

- 13.1 Overview of Mechanical Modeling — 174
  - System of units — 176
  - Implementing mechanical models — 177
- 13.2 Automotive Seat Position Controller — 179
  - Mechanical models — 179
  - Modeling the temperature-limited MOSFET — 182
  - Integrated seat position system simulation — 182
  - Conclusions — 184
- 13.3 References — 184

14. Motors — 185

- 14.1 Overview of Motor Modeling — 186
  - Implementing Motor Models — 186
- 14.2 Floppy Disk Drive Head Position Controller — 187
  - Modeling of stepper motor driver ASIC — 189
  - Modeling of stepper motor — 190
  - Modeling the mechanical head positioning system — 192
  - Simulation results — 193
  - Conclusions — 195
- 14.3 References — 198

15. Hydraulics — 199

- 15.1 Overview — 199
- 15.2 Fundamental Hydraulic Quantities — 200
  - Flexible transmission line model — 200
- 15.3 Example Design—Simple Hydraulic Press — 202
  - Simulation results — 202
- 15.4 References — 204

16. Electro-thermal Systems — 205

- 16.1 Electro-thermal Network Modeling and Simulation — 206
  - Electro-thermal modeling and simulation — 207

		Electro-thermal semiconductor models	209
		Thermal network models	212
	16.2	Electro-Thermal Network for a PWM Inverter	214
		Circuit topology and models	214
		Simulation results	217
		Conclusions	220
	16.3	References	221
17.	**Magnetics**		**223**
	17.1	Overview	223
		Electromagnetics	224
		The inductor	225
	17.2	Fundamental Magnetic Quantities	225
		Magnetomotive force (mmf)	226
		Flux (Φ)	227
		Flux density (B)	228
		Permeability (μ)	228
		Magnetic field strength (H)	229
	17.3	Nonlinear Magnetics	229
		Winding model	230
	17.4	Power Supply With Electromagnetic Models	232
		Simulation results	233
	17.5	References	236
18.	**Summary**		**237**
	18.1	References	238
Appendix A.		**Reference Information**	**239**
	A.1	MAST Reference	239
		General syntax	239
		Connection points (pins)	240
		Include files	241
		Unit definitions	241
		Pin definitions	242
		Parameters	242
		Parameter data types	243

		Expressions	244
		Unary and binary operators	245
		Intrinsic functions	245
		Declaration operators	247
		Assignment statements	248
		If constructs (condition checking)	249
		Simulator variables	250
		Foreign functions	251
		MAST functions	251
	A.2	Newton Steps	252
		Specifying newton steps	253
		Example	255
	A.3	References	256

Appendix B. MAST Template and Function Listings — **257**

	Chapter 4—Behavioral deadband	257
	Chapter 4—Netlist of final design (Figure 4-8a)	258
	Chapter 6—Behavioral pole-zero transfer function	259
	Chapter 9—MOSFET MAST function mosparam	260
	Chapter 9—MOSFET MAST function mosval	262
	Chapter 9—MOSFET MAST function mosdiode	265

Index — **267**

Preface

Buckminster Fuller defined synergy as *"the unexpected interaction of parts in combination...wholes always contain behaviors you couldn't have expected when the parts were strewn in front of you."* Simulation is a tool that a designer can use to maximize the benefits of synergy and minimize its pitfalls (i.e., to avoid Murphy's Law).

Simulation has been around for a long time. In fact, scale replicas ("models") have been used since antiquity to determine the interaction of parts before committing resources to major projects. One of the first uses of digital computers was to perform the calculations and store the data of simulation. Before that, analog computers helped manage the vagaries of design.

Despite this long history of simulation, fundamental challenges still remain:

- Designs analyzed using simulation are challenging by nature—there's little need to simulate a simple design.
- Simulation is abstract by nature.

- Learning the tools and techniques of simulation can be arduous. There are many things to keep track of—complexities of the design, how to use the simulator, debugging the design, and negotiating the limitations of the simulator.

Historically, one of the greatest limitations of simulators has been their intended scope. For example, a number of simulators (i.e., SPICE and its derivatives) have been written with an emphasis on small analog integrated circuits (ICs). As long as the designer is working in this realm, such a simulator is a very valuable tool. However, the further the design strays from the intended scope of the simulator, the fit (and the designer) becomes increasingly tenuous. Even though this type of specialized simulator may be able to simulate a large analog IC, *it will invariably lose accuracy and ease-of-use* on sections of the design that are at a level of abstraction higher than transistors or op amps. This restricted realm of operation prohibits the designer from using the top-down techniques necessary to master the complexity of large systems. Of course, this phenomenon is compounded as the designer moves on to a system of which an IC is only a part.

As a further example, consider an IC that controls a system of electrical and nonelectrical components (such as in a CD player, fuel injector, or hydraulic lift). For a simulator that is principally intended for analog ICs, there will come a point at which both the nonelectrical components and the overall system must be omitted from the simulation. This not only prevents the designer from capitalizing on the benefits of synergy, but it can also signify that the tool is no longer useful—the inevitable consequence of being limited to simulation at the transistor level.

Why an AHDL?

A simulation can be performed only on models that are usable by the simulator. For this reason, the models determine the level of abstraction that can be achieved. Moreover, with the tremendous increase of available devices and transducers that link previously disparate technologies, it is no longer possible (nor desirable) to incorporate all the models a designer may need *as part of the simulation program*. The ability to create models, therefore, is critical.

The Saber simulator from Analogy, Inc. was developed in 1986 as the first commercial simulator to separate the models from the simulation program. With this development came the first analog hardware description language (AHDL), MAST. A hardware description language, such as the MAST[1] modeling language, enables designers to create new models and combine them with existing models to investigate their synergistic behaviors within the same design. In fact, the capabilities of MAST have been expanded to permit creation of mixed signal (analog-digital) models. The alternative to using a language to create models outside the simulation program is to incorporate disparate models into the simulator itself. Over the years, this has proven to be an arduous undertaking with little likelihood of success.

Although this book will attempt to abide by this separation of simulation and modeling as much as possible, there will be occasions where the distinction blurs. This is not a fault of either domain, but it is an inevitable consequence of their interaction. It is not the intent of the authors to eliminate such discussions, but rather to provide instructional context with minimal digression, to ensure that the focus stays directed to the modeling topic at hand.

Goals of this book

The goals of this book are stated below and are predicated on the assumption of the usefulness of computer simulation:

- To establish the importance of using an AHDL to make computer simulation more powerful, efficient, and valuable.

- To provide the understanding of what is needed to develop a model using an AHDL.

That is, this book spells out, in general terms, what modeling with an AHDL adds to the existing field of computer simulation and then uses specific examples to develop that understanding. Although this book uses the MAST modeling language to achieve the goals listed above, it is not intended either to teach model theory or to be an exhaustive reference on the syntax of the language. Consequently, model implementations are provided in a straightforward manner and without justification of all their details.

---

[1] MAST is not an acronym.

Further, although other hardware description languages have been recently introduced (such as HDL-A™ and SpectreHDL™), there has not been sufficient opportunity for them to be exercised with their particular simulators nor for applications to be written and tested. The MAST language, because it is a mature product, meets both of these criteria, which are in turn indispensable for meeting the goals of this book.

Intended audience

Because of the growing importance and use of simulation and modeling, there is some reluctance to define the intended audience too narrowly. However, those readers who will get the most from this book are engineering students and practicing engineers who are interested in familiarizing themselves with the characteristics of an AHDL, as stated above for the goals of this book.

Of course, a prospective modeler must still fulfill the requirements for creating models, such as knowing the appropriate equations and algorithms with which to represent them. For the student or working professional, much of this information is available in textbooks. If a model is complex, it will be necessary to consult a reference for the intricacies of the modeling language. Nonetheless, in the coming years, engineers will find it necessary to understand how to design using an AHDL.

Moreover, if the contents of this book bear a resemblance to product documentation at times, this is done only to help bridge the gap between such references and academic textbooks. That is, the actual product documentation for the MAST language provides much more thorough coverage of the details of writing models than can be addressed by this book.[1]

The chapters of this book are divided into three major sections:

- *Fundamentals of Modeling* provides an overview of general modeling and simulation concepts that are used in subsequent chapters. These introductory chapters cover topics such as macromodels, behavioral models, primitive device

---

[1] As of this writing, there is also an effort underway to produce a Language Reference Manual for the analog extension to VHDL—VHSIC Hardware Description Language (IEEE Standard 1076.1).

models, modeling hierarchy, top-down design, non-electrical technologies, and the Newton-Raphson iterative simulation technique. These topics are presented to help further the understanding of what is needed to develop models in an AHDL.

- *Model Implementation* begins to convey the implementation details of the MAST AHDL. The chapters in this section show how to use the governing equations of several commonly used models, along with equations that are readily available from well-known textbooks and papers. This information is provided in both tutorial and reference fashion, serving as an introduction to the basics of the MAST AHDL. Each chapter builds on the information from preceding chapters in order to demonstrate progressively more complex modeling concepts. This culminates with the diode and MOSFET models given in Chapter 9, which are intended to show the depth of the MAST language and which may be of interest to a more specialized segment of the modeling population.

- *Advanced Applications* contains several examples of designs that use models written in the MAST AHDL. Each example makes use of concepts brought up in the first two sections. The main purpose of these chapters is to illustrate the importance of using an AHDL to enhance the power of computer simulation.

For the novice modeler, this book should prove readily adaptable to the immediate task at hand. The examples presented in this book have been developed and performed using the Saber simulator. They can be used as is to gain a feel for the nature of the language and the various facets of modeling. As such, the reader should be able to sit down at a computer that has Saber installed, take this book and the appropriate texts, and adapt the examples to the equations and algorithms in those texts to models in the limited time available to both students and practicing engineers.

Acknowledgments

The authors would like to express their appreciation and gratitude to the many persons who were essential to the realization of this book. In particular, we wish to acknowledge the material contributions of Lewis Sternberg, Ernst Christen, Darrell Teegarden, Mike Donnelly, Genhong Ruan, and Jeff Carlson. We also want to thank Ian Getreu, Ernst Christen, Genhong Ruan, Chris Wolff, Doug Johnson, and Nic Herriges for their reviewing efforts, along with Laura Churchill and Kimberly Keyes for their assistance in production and quality assurance. In addition, we would like to thank the entire staff at Analogy, Inc., for their support of this project in particular and for their product development efforts in general. Specifically, we thank Ian Getreu for his pioneering vision in developing MAST models and establishing libraries, David Smith for his ability to make a simulator and a modeling language work together as a viable product, and Martin Vlach, without whom there would be no MAST AHDL about which to write a book.

Finally, we thank our abiding wives, Mary and Diane, and our families for their (nearly) unlimited patience and support.

H. Alan Mantooth
Mike Fiegenbaum

August 1994

Section 1

Fundamentals of Modeling

Chapters 1-4

This section provides an overview of general modeling and simulation concepts that are used in Sections 2 and 3. Chapters 1-4 cover topics such as macromodels, behavioral models, primitive device models, modeling hierarchy, top-down design, non-electrical technologies, and the Newton-Raphson iterative simulation technique. These topics are presented to help further the understanding of what is needed to develop models in an AHDL.

CHAPTER 1

Modeling: A Working Definition

This chapter describes the issues concerning modeling and simulation. It begins with a brief description of classical continuous-time simulation techniques and concludes by establishing the context within which models are used.

1.1 What Is Simulation?

There are some excellent texts devoted to the topic of continuous-time numerical simulation [1, 2]. It is beyond the scope of this book to present all of the details associated with the algorithms that may be employed to solve ordinary differential equations (ODEs) numerically. However, it is necessary to summarize the basics of this material in order to provide understanding of what a simulator does and to reduce the mystery of a simulator as a "black box." This understanding of the simulation process serves as a starting point for a discussion of what is required to develop models. Moreover, it is essential if these models are to be computer efficient and numerically robust.

For this discussion, the simulation process is one in which a system of nonlinear ordinary differential equations is solved. Typically, in circuit simulators, these differential equations are not input directly, but are derived from each of the models that are interconnected in the netlist (*net*work *list*ing). The circuit or system is described to the simulator in the form of a netlist, which indicates how the various devices are connected in a circuit. Each device model has an associated matrix stamp or matrix of equations that is used within the overall circuit (system) matrix to be solved by the simulator. Once this circuit matrix of equations is formulated, it is ready to be solved. As a simulation analysis (such as a transient analysis) proceeds, the system matrix is updated with new values from the models.

Figure 1-1a depicts the implementation of the simulation process in detail. As a means to a solution, the simulator provides an initial "guess" for the values of the independent variables, such as node voltages in a circuit, which is refined at successive iterations in the simulation. The simulator solves the system of equations based on the returned values of the dependent variables from the models, subject to the condition, for electrical circuits, that all currents sum to zero at a node (which is known as Kirchhoff's Current Law or KCL).

Figure 1-1b shows a block diagram of how a model fits into the overall simulation process. Essentially, an AHDL model provides an efficient mechanism for relating the independent and dependent variables so they can be used by the simulator.

Each model equation is formulated so that a dependent variable is expressed in terms of independent variables. Confining this description to electrical circuits, the independent variables can be chosen as current, voltage, charge or whatever is convenient. However, the simulator imposes a specific conservative condition on current (KCL) when solving the system of equations. This notion of a conservative system can be generalized by referring to two entities: *through variables* and *across variables*. For electrical circuits, the through variable is current and the across variable is voltage. As other technologies are described in later chapters the respective through and across variables will be identified. Through and across variables are important to understand because they indicate to the simulator the physical quantities to

Modeling: A Working Definition

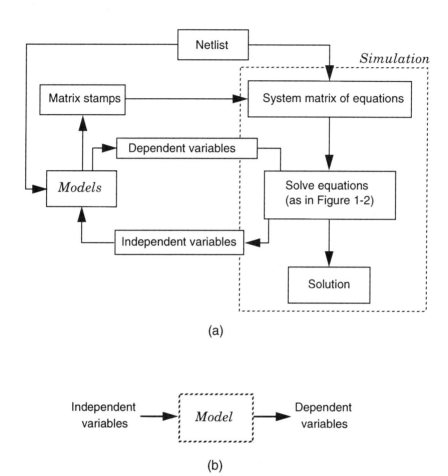

Figure 1-1. *The continuous-time simulation process (a) implementation (b) overview.*

be conserved (e.g., adhering to KCL or KVL). An AHDL simulation model associates a through and an across variable with each connection point (or node) in the model. In general, the through variables are summed to zero at a connection point as a means to insure a conservative system (another statement of KCL).

Figure 1-2 [3] depicts the process of solving ODEs into a series of steps:

1. The most general starting point is the transient solution of the nonlinear ordinary differential equations. Other analyses (e.g., linear frequency domain) are a subset of this process.
2. Numerical integration techniques (such as Gear or trapezoidal) are first employed to transform the given system of ordinary differential equations into a series of nonlinear algebraic equations [4].
3. An iterative technique, typically Newton-Raphson, is used to transform the nonlinear algebraic equations into a series of linear algebraic equations.
4. The system of linear equations can now be solved using direct matrix techniques (such as LU decomposition).

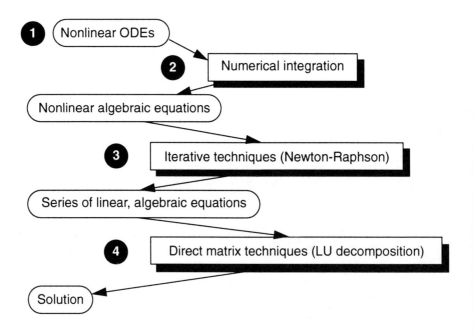

Figure 1-2. Numerical solution of ODEs.

Most commercial simulation programs (such as SPICE [5], its derivatives, and the Saber simulator) employ the simulation process shown in Figures 1-1 and 1-2. These depictions of the simulation process illustrate how circuits are simulated and where models come into play in this process. For the purpose of this book, knowing the specific details of this process is not nearly as important as having a more general understanding of how the simulation process is performed and how models fit into it.

Simulation and modeling: the user's perspective

When first confronted with computer simulation, most designers in the field of electrical engineering have the natural impression that a simulator is essentially a software version of an oscilloscope or logic analyzer. While this conceptualization can be quite useful, it does have limitations. To get the most out of simulation, a new viewpoint is needed.

Simulation is a technique by which a user asks a simulation program questions and it returns answers. A simulation program is an investigative tool, and as such it is indeed similar to an oscilloscope or a logic analyzer (when they are used well). Simulation, however, allows one to ask more direct and insightful questions. Of course, the quality of the answers depends upon the quality of the questions. A simulation can be performed only on a model that provides data in a form that is usable to the simulator. Consequently, models are an implicit part of the question. Models may limit the questions that can be asked; they may also limit or enhance the answers delivered.

For example, a commonly used operational amplifier model in the simulation industry is the Boyle model [6]. Given the appropriate parameter values, this model accurately predicts many critical frequency effects and nonlinearities. Many important and useful questions may be answered—as long as the model is exercised in its intended region of operation. However, one of the behaviors not predicted by the standard Boyle model is nonlinear output impedance as a function of current and frequency. This model cannot accurately simulate the behavior of an op amp driving a finite load at high frequencies.

Similarly, linear inductor models can be very useful, as long as the actual circuit is not affected by the frequency-dependent hysteresis of an actual inductor. A simple resistor can model a lamp reasonably well, as long as the circuit is not affected by inrush current when the lamp is turned on.

Models can also affect how easy it is to ask questions, and how easy it is to get and interpret the answers. A mechanical system can be represented by existing models for electrical components by letting a spring, mass, and dashpot correspond to inductor, capacitor, and resistor models. However, the answers must be translated from currents and voltages to forces and speeds. ("Now over here, I is the torque on the rotor, and over there, V is the heat from friction...or is it the other way around?") On the other hand, a model can make life easier by supplying answers without being prompted. For example, if a resistor model is exercised beyond its Safe Operating Area (SOA), it may automatically generate data for a report, issue a warning, and alter its own value—allowing one to see how the circuit responds to such a failure. ("Are those $10 transistors really protected?")

Of course, models must have a simulator with which they can be exercised efficiently and without problems such as convergence and discontinuities. ("What do you mean this switch model doesn't work? It's mathematically correct!")

1.2 What Is a "Model" Anyway?

This is a good question that is best answered by examining a sequence of model categories.

1. A *conceptual model* is a mathematical, algorithmic, or pictorial representation of a physical thing or concept. Typically, a conceptual model will use parameters to provide the same equation for multiple instances. For example, Newton's law of gravity is generalized to describe the gravitational force between any two bodies, given the proper values: the mass of each of the bodies and the distance between them.

   Figure 1-3 illustrates some examples of conceptual models.

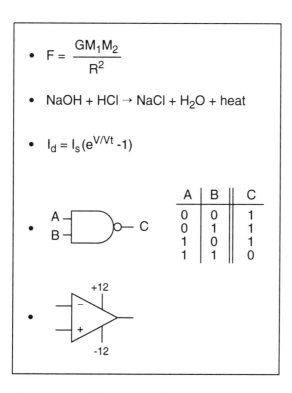

Figure 1-3. Examples of conceptual models.

2. A *generic model* is an encoded (programmed) representation of a conceptual model. It is this type of model that is principally addressed by this book. Such models may be written in an AHDL (such as MAST) or in a programming language such as C or FORTRAN. Incidentally, models written in FORTRAN or C can be called directly from MAST.

A simple diode model written in MAST (which is referred to as a *template*) is shown below.

```
template diode p n = Is, Vt
electrical p, n
number Is, Vt          # Is and Vt are the parameters for this simple diode
                       # model (saturation current and thermal voltage)
{
    branch id = i(p->n), vd = v(p,n)
    id = Is * (exp(vd/Vt) - 1)
}
```

3. A *component model* is the result of combining a set of specific parameter values with a generic model. In MAST, this is referred to as a component template. Because traditional simulators do not allow for the creation of generic models, many simulator users are only familiar with component models. The process of determining parameter values for a model is called *characterization*. For example, values for Is and Vt could be specified for the diode template above that match those published for the commercially available 1N198 diode.

When creating models for simulation, this sequence of models should be in harmony with each other and with the questions that arise about any design that incorporates them. A model that cannot be coded and efficiently simulated has limited utility. Similarly, during creation of the generic model, characterization must be considered. If the parameters of the generic model are not orthogonal, or if they are difficult to measure, characterization will be hampered.

A MOSFET is a MOSFET is a MOSFET...or is it?

As an example, consider the Metal Oxide on Silicon Field Effect Transistor—the MOSFET. The physics behind many MOSFET models is pretty much the same, with some sign changes to account for enhancement or depletion mode devices and n-type or p-type conduction paths. However, different users may ask for very different, and sometimes conflicting, models of a MOSFET.

Modeling: A Working Definition

On one hand, *a MOSFET designer* will be interested in how various geometries, doping levels, and junction depths affect the behavior of a device. A very detailed, three-dimensional model is required for this, taking into account the local fields. Several dozen model parameters will be provided for geometry and processes. The designer can afford to use a very detailed model because only one or two transistors will be simulated at a time.

On the other hand, this model would not be appropriate to a *circuit designer*. The parameter values would be too difficult to determine without intimate (and probably proprietary) knowledge of the fabrication process. Even if the parameter values were known, the simulation times for even a medium-sized circuit (say about 100 elements) would be prohibitive.

A circuit designer will be concerned with much different effects if the MOSFET functions principally as an analog switch in a "glue" chip between analog and digital circuits. It is quite likely that the model of an ideal switch with appropriate threshold values, finite impedances and switching times will be sufficient. As long as such chips are operated within their intended region of operation, the fact that a MOSFET model is used for this switch can be ignored. Model parameters will be provided in terms of commonly available data book specifications. Moreover, simulation results can be verified against the data book specifications if they are available from the manufacturer. Verification against laboratory measurements of the physical device may be necessary for critical designs.

Furthermore, the same circuit designer, when creating a switched capacitor filter or a discrete amplifier with MOSFETs, will need a detailed model that takes into account other characteristics, such as nonlinear capacitances, short-channel effects, charge conservation, and temperature effects, to name a few. For such components, (expensive) laboratory characterization is necessary. Data books do not always provide all the critical parameters. If many such components require characterization, then the model should also be designed to ensure that this can be done quickly and reliably.

Obviously, a designer working with *power* MOSFETs will be concerned with different effects, such as the pinchoff effect in vertical power MOSFET structures.

This description of MOSFETs demonstrates the range of focus available to a modeler. One major objective of this book is to provide understanding of how a modeling language—an AHDL—is capable of describing hardware behavior over such a range. An AHDL and the simulator with which it is used may eliminate most of the issues involved with coding a new model. Nonetheless, it is still incumbent upon the model writer to evaluate which characteristics are to be included in the model and how they will be evaluated, regardless of how the model is implemented or the simulator being used.

1.3 References

1. J. Vlach and K. Singhal, *Computer Methods for Circuit Analysis and Design*, Van Nostrand Reinhold, New York, NY, 1983.
2. L. O. Chua and P.-M. Lin, *Computer-Aided Analysis of Electronic Circuits: Algorithms and Computational Techniques*, Prentice-Hall, Englewood Cliffs, New Jersey, 1975.
3. H. A. Mantooth and M. Vlach, "Beyond SPICE with Saber and MAST," *IEEE Proc. of Int. Symposium on Circuits Syst.*, Vol. 1, pp. 77-80, May 1992.
4. K. E. Atkinson, *An Introduction to Numerical Analysis*, John Wiley & Sons, New York, NY, 1978.
5. L. W. Nagel and D. O. Pederson, "Simulation program with integrated circuit emphasis," ERL Memo No. ERL-M520, University of California, Berkeley, May 1975.
6. G. R. Boyle, B. R. Cohn, D. O. Pederson, and J. E. Solomon, "Macromodeling of Integrated Circuit Operational Amplifiers," *IEEE J. Solid-State Circuits*, Vol. SC-9, pp. 353-363, Dec. 1974.

CHAPTER 2

Modeling With an Analog HDL

This chapter begins discussion of how a hardware description language (HDL) contributes to modeling and simulation. This includes many important universal modeling concepts and how these are readily accomplished using an HDL.

2.1 What Is an AHDL?

An analog HDL is a programming language that is specifically designed to allow the description (or modeling) of hardware that performs a continuous-time function. An analog HDL must possess the necessary constructs that enable this description to be used in the traditional domains of simulation and analysis—principally the time domain and the frequency domain. As it turns out, it is also necessary to consider an initialization domain or "time zero" as a separate domain, which is commonly referred to as the DC domain. This is in contrast to a digital HDL (such as VHDL) where only two domains are of interest: the initialization domain and the time domain. Another important contrast to a digital HDL is that, by definition, an AHDL is designed to operate in conjunction with a continuous-time simulator (such as SPICE or Saber) as described in Chapter 1.

Note, however, that an analog HDL need not be limited to only continuous-time phenomena. It is very useful and powerful to have the ability to describe behavior that may either be discrete in time and/or discrete in value (e.g., sample-and-hold). These "events" in time or value are more consistent with the digital domain. This approach of modeling events is known as *event-driven modeling*. Because MAST has this capability also, it is actually a mixed-signal (analog-digital) HDL (this is described in more detail in Chapters 10 and 11).

An example of event-driven modeling is the voltage comparator shown in Figure 2-1. The two inputs on the left of the device are analog pins that are continuously monitored; the output on the right is a 4-state (1, 0, X, Z) digital output. The inputs can have analog functionality, such as offset and hysteresis, while still producing a digital output.

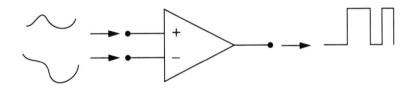

Figure 2-1. Voltage comparator.

2.2 Modeling Concepts

Mathematics

At the heart of any discussion of an AHDL is the most fundamental concept of modeling—mathematics. An analog HDL, together with an appropriate simulator, should be able to model any system that can be described with nonlinear, ordinary differential equations (ODEs). However, this statement doesn't quite capture the entire picture. In all technologies, there are fundamental laws that are enforced by the simulator, regardless of technology or complexity. In electrical circuits, such laws are Kirchhoff's current law (KCL) and Kirchhoff's voltage law (KVL). There are also constraint relationships that may or may not be included in a model but that the simulator does not enforce. Conservation of charge in a MOSFET model is an example of such a constraint. There are many MOSFET models that do *not* conserve charge, while there are some that do. Whether the constraint is applied or not depends on the implementation of the model.

For electrical circuits, KCL is enforced by the simulator in the process of finding a valid solution to the system of equations. As mentioned in Chapter 1, a continuous-time simulation program solves a system of simultaneous ODEs that are formulated so that the dependent variable is expressed in terms of independent variables. To comply with this, the analog HDL provides the concept of *through* and *across* variables, which are used by the simulator as the dependent and independent variables of the ODE at a node. The through variable at a node (electrical current, dissipative power, and hydraulic flow) is summed to zero at a node by the simulator. The across variable (electrical voltage, temperature, and hydraulic pressure) is typically taken as the independent variable that the simulator iteratively refines from the initial guess to solve the system equations.

Another aspect of solving differential equations is in providing for the expression of implicit equations—when the roles of through and across variables are reversed (i.e., the through variable is an independent variable and the across variable is a dependent variable). These types of equations must be considered simultaneously with the other system equations being solved for by the simulator. Thus, the *system variable* is the across or through variable taken as the independent variable for a given system of equations.

Finally, an analog HDL must be able to implement initial conditions for these equations to be solved correctly. Examples of these concepts will be provided in later chapters.

Conceptualizing a model

For the moment, consider only the continuous-time aspect of an analog HDL. With such a language, the modeler can express hardware behavior in terms of nonlinear, ordinary differential equations (ODEs). Continuous-time simulation programs such as SPICE or the Saber simulator solve nonlinear ODEs numerically.

The task of using a modeling language most effectively becomes one of connecting a model to the AHDL, which consists primarily of posing the system of nonlinear ODEs to the simulator for numerical solution. In general, there are several thought processes required to set up these equations. These are described below, but the order in which they are used may vary depending on the complexity and availability of information on the model being created.

1. The modeler must determine how the model to be created will interface to other models (i.e., define inputs and outputs).
2. At some point the ODEs must be ordered or related to one another in a way that conveys the desired outputs based on the specified inputs. This relationship is often referred to as the *topology* of the model. It can be loosely thought of as the large-signal model.
3. Next, the algebraic relationships that are required to describe the variables in the differential equations must be determined.
4. Finally, the model parameters must be chosen.

These thought processes are necessary to create even the simplest model. Each one partitions the description of how the model uses input variables and model parameters to determine the various coefficients of the differential equations that are in turn solved by the simulator. After the simulator calculates the solution of those equations, they are provided as outputs of the model.

If the digital (event-driven) domain of the analog HDL is taken into consideration, the only thought process that changes from that described above is in setting up the coefficients in the differential equations. These coefficients can be made functions of

Modeling With an Analog HDL 17

discrete-time or discrete-valued entities (or both). However, the system of equations being solved remain nonlinear ODEs. The model parameterization issue is also the same for both analog and event-driven.

Figure 2-2 shows parallel block diagrams that depict the modeling thought process (on the left) and a general implementation process (on the right). Chapter 5 provides a more implementation-specific explanation of the correspondence between the left and right blocks. In addition, the examples of AHDL models in

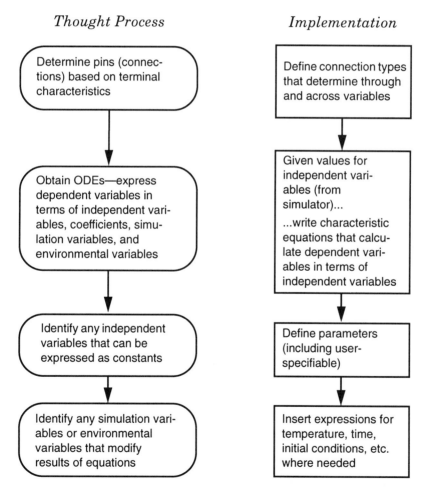

Figure 2-2. Model implementation.

subsequent chapters provide further elaboration on techniques of converting model information to model implementation.

2.3 What Can Really be done with an Analog HDL?

The following sections describe some traditional areas of modeling to which the MAST AHDL has proven to be particularly well-suited.

Other technologies

Most of the discussion thus far has referred to, in one way or another, electrical circuits as the subject of modeling. However, an analog HDL easily allows the implementation of models across a variety of technologies. The analog modeling hierarchy is equally applicable to non-electrical systems.

The ability to apply the fundamental laws of a technology such as conservation of energy, Newton's laws, etc. in the "native tongue" of that discipline is very powerful. Electrical engineers have learned that, when designing mixed analog-digital circuits, many of the problems encountered in these chips arise in the interface between analog and digital circuits. This is the reason that mixed analog-digital simulation has become and remains such a serious need in the EDA industry. The ability to model and focus on the analog-digital interface has proven to be extremely valuable in finding design errors.

The same value is realized when the interface between non-electrical technologies is modeled. Section 3 (Chapters 13-17) of this text provides some examples of these types of systems and effects, such as electro-mechanical, electro-thermal, and electro-magnetic.

Modeling hierarchy

Table 2-1 shows three general levels of models used in analog simulation: primitive, functional, and behavioral. With an AHDL, one can represent a model at any level of this hierarchy, as well as connect models from different levels within the same design. This capability to freely traverse the modeling hierarchy is one of the most powerful features of an AHDL.

Table 2-1. Analog modeling hierarchy.

Design abstraction	Modeling characteristics
Primitive	Devices (MOS, BJT, diode, etc.) represented by: - Analytical equations - Tables
Functional	Macromodels derived by: - Circuit simplification - Circuit build-up - Symbolic methods - Combinations of the above
Behavioral	High Level language descriptions - Linear and nonlinear mathematical equations - Tables

- The *primitive* level is the lowest level. The elements used to represent the system are ones of the finest level of detail practical for this type of simulation. For electrical circuits, this is typically the transistor level. The circuit is represented by the semiconductor devices (such as a bipolar junction transistor [1]) and passive devices (such as resistors and capacitors) that are actually in the circuit.

- The *functional* level is traditionally considered as consisting of macromodels that are collections of primitive devices "designed" to model specific functionality. Macromodels such as that of an operational amplifier usually possess some correlation to the original circuit topology that they are intending to model (see Chapter 3).

- The *behavioral* level, as the name implies, provides models for arbitrary devices, functions, or blocks. Such a model can consist of equations or procedural descriptions that may not correspond to any single circuit element (such as a transfer function). However, it would be incorrect to think of behavioral models as simply idealizations or linearizations. They can be as simple or complex as required, with as much accuracy as any corresponding primitive. This relationship of hierarchical levels is covered in Chapter 6.

Macromodeling

Macromodeling is a popular and useful way to combine existing models to produce another model. The Boyle op amp model mentioned in Chapter 1 is a classic example of macromodeling. Because of its circuit design nature, macromodeling has found great favor with designers—it is much like "playing with circuits." Although it might seem likely that proponents of an HDL would be seeking to change this approach to modeling, the opposite is true. Macromodeling remains a powerful concept and should be taken advantage of wherever appropriate. However, a primary disadvantage of macromodeling is that the technique depends on the availability of underlying models. If these are limited, then the macromodeling approach and what it can ultimately achieve are limited. What AHDL proponents do favor is adding to the building blocks that are available to perform this task. Instead of working with a limited inventory of controlled sources, transistors, diodes, resistors, capacitors and inductors, why not have arbitrary building blocks from which entirely new varieties of macromodels can be created? The analog HDL provides this capability. Macromodeling is covered more thoroughly in Chapter 3.

Behavioral modeling

While macromodeling is a viable modeling technique, it can sometimes be far too unwieldy to create meaningful models. Often, a simple behavioral model is far more efficient and possibly more accurate due to the directness of the implementation. An excellent example of this is the comparison of methods describing ordinary differential equations, provided in Chapter 3.

As a simple example, consider the task of modeling the systematic offset voltage of an operational amplifier or a voltage comparator as a function of temperature. In reality, this non-ideality arises due to the mismatch between transistors in the circuit or to asymmetry in the circuit topology. In order to accurately predict the offset voltage over temperature, a designer would not only have to properly account for these mismatches, but also for tracking these mismatches over temperature. However, it is a simple matter for a behavioral model to include an equation that introduces an offset that is a function of temperature. Measured data could easily be fit to such an equation very accurately (i.e., it would be simpler and computationally faster).

Semiconductor device modeling

Semiconductor device models of all types have been implemented in MAST, reflecting the full range of complexity in the industry. Empirical, semi-empirical, and physical models for MOSFETs, diodes, bipolar transistors, GaAs MESFETs, and HEMTs (to name a few) have been implemented. The physical models fall into one of two categories:

- Charge-control models
- Discretized partial differential equation-based (PDE-based) models

The charge-control models are similar to those originally implemented in SPICE. These models attempt to describe device behavior by modeling the movement of charge through the various physical regions of the device. Basic charge-control models have proven extremely useful; with the addition of nonquasi-static effects, these models have been improved to apply to higher frequencies as well.

The discretized PDE-based models really address a different problem. These models essentially discretize the regions within the device to more closely approximate the dynamic movement of charge in the device. This gives the model more of a finite element "feel." However, a discretized PDE-based model is useful for characterizing fabrication processes and modeling high-current, high-power effects such as current filamentation. This modeling approach has been applied to such devices as silicon-controlled rectifiers (SCRs), gate turn-off thyristors (GTOs), power MOSFETs, power diodes, and IGBTs.

An example of a physics-based analytical model that has been implemented in MAST is the power diode [2, 3]. The version of this model that is generally available contains many effects that are not available in the traditional SPICE diode model. The

MAST power diode model has an extensive DC representation for the diode current as a function of terminal voltage. Its parameters allow characterization of such forward effects as low-level recombination in the depletion region, high-level injection, emitter recombination, and series resistance. Figure 2-3 shows these effects, which have been exaggerated for the purpose of illustration.

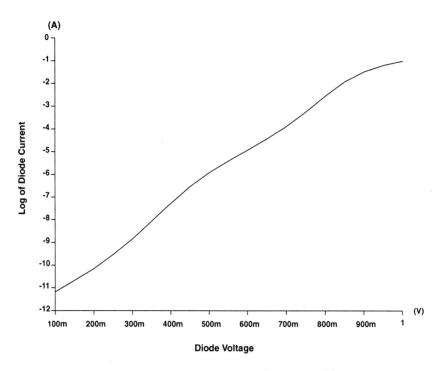

Figure 2-3.   I-V curve of a power diode.

In the reverse region of operation, the model is capable of predicting exponentially varying leakage currents, as opposed to the constant leakage found in other models. This is an important effect in some zener diodes that exhibit a "soft knee" characteristic [4]. Further, the power diode model can be characterized to predict avalanche or zener breakdown. Finally, the model is capable of exhibiting a different series resistance in the reverse region of operation than that in the forward region.

The transient effects included in the model are equally numerous. The model is capable of predicting forward and reverse recovery, as shown in Figures 2-4 and 2-5. The reverse recovery can be made a function of bias as well. The model has complete temperature effects including self-heating if desired [5, 6].

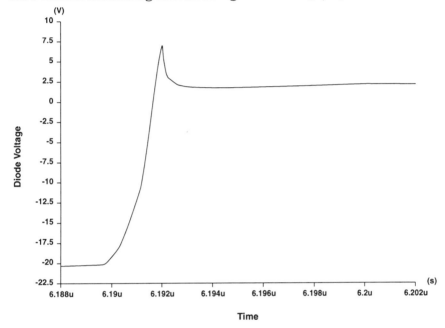

Figure 2-4.   Power diode—forward recovery.

This example indicates the level of detail that can be included in a model with an analog HDL. This power diode model consists of implicit relationships and nonlinearities that would be very difficult to capture in a macromodel.

Model validation

No matter which technique is used to implement a model, or which choice is made between a macromodel or a behavioral model, it is always necessary to validate the model through simulation. For models of semiconductor devices, this involves comparisons between experimental measurements and simulation results from the model. For macromodels and behavioral models, the comparison is typically done either to the original circuit or to

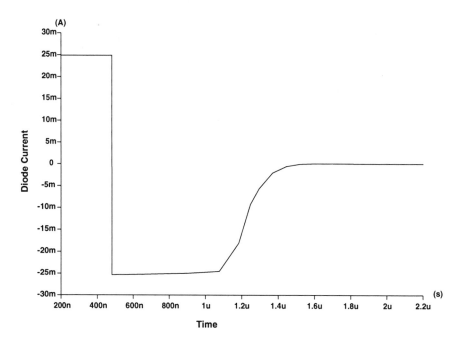

Figure 2-5. Power diode—reverse recovery.

published data sheet information. However, in some cases, experimental data is also available for validation.

There are basically two levels of validation. The first is *feature validation*. This involves subjecting the model to a test condition in order to determine if it is capable of modeling a given effect. The second level is how accurately the model can predict a given effect, assuming that it is present. This is referred to as *accuracy validation*.

Feature validation is usually quite straightforward. Accuracy validation can require more experience and research. While a model may predict a given effect, answering the question of how well it predicts this effect is a different matter. Consider the reverse recovery effect of the power diode presented previously. Simulation of a simple test circuit determines whether or not the model can predict this effect. However, the effect is known to vary among different device structures and with bias conditions and temperature. Validating these effects is a very involved process. Constrained optimization and parameter extraction play key roles in validating the model over such varied conditions.

Standardization

Establishing a standard analog hardware description language will eventually force EDA companies that provide an analog HDL to provide a minimum of functionality in their language and, in turn, their simulator. The major benefit of this is that a model that conforms to the standard can be run on different simulation programs without alteration. To users of an analog HDL, this is the single most important advantage of having a standard.

However, having a standard language does not guarantee "standard results." Since the simulation algorithms are not standardized, different simulators will in general produce different results with the same model. The same model may behave quite differently from a convergence standpoint as well. The reality is that adjustments will be required when standardized models are used on different simulators. These adjustments may have to be made to the model and/or the simulator algorithm settings in order to achieve standard results.

Another advantage of having a standard is that it increases the availability of models. This means there are more building blocks available for modeling, which reduces the amount of time that a designer has to spend developing models. This leads to more available time to spend on the actual design. The establishment of a standard also removes much of the risk associated with large model development tasks by eliminating the possibility that the model will not be usable in the future.

Standardization does not come without a price. The standard will be the least common denominator of the set of features required to model analog behavior. One primary disadvantage of a language standard is that development of advanced language features will be slowed. As more resources are devoted to standardization compliance, less will be available for new feature development. Further, the return on the investment of providing additional features has to be more carefully understood, since these features are not part of the standard (i.e., as needed features are added as extensions to the standard, the updated language is no longer a standard). This is a likely scenario, since EDA vendors will continue to pursue market differentiation for competitive reasons.

The question that simulation users will face is "Will the differentiation created by these new features in the language justify moving away from the standard?"

2.4 References

1. I. E. Getreu, *Modeling the Bipolar Transistor*, Tektronix, Inc., 1976.
2. C. L. Ma and P. O. Lauritzen, "A simple power diode model with forward and reverse recovery," *IEEE Transactions on Power Electronics*, Vol. 8, pp. 342-346, Oct. 1993.
3. Power diode template description, *Standard Template Library* manual, Analogy, Inc., May 1994.
4. A. H. Pawlikiewicz and G. Zack, "New macromodel for zeners," *IEEE Circuits & Devices: Simulation and Modeling Column*, Vol. 9, pp. 7-11, Mar. 1993.
5. H. A. Mantooth and A. R. Hefner, "Electro-thermal Simulation of an IGBT PWM Inverter," *IEEE Proceedings of PESC '93*, pp. 75-84, Seattle, Washington, June 1993.
6. A. R. Hefner and D. L. Blackburn, "Simulating the dynamic electro-thermal behavior of power electronic circuits and systems," *IEEE Transactions on Power Electronics*, Vol. 8, pp. 376-385, Oct. 1993.

CHAPTER 3

Macromodeling

A macromodel is a simplified equivalent circuit or a simplified system of equations used to represent the terminal characteristics of a circuit or system. The goal of macromodeling is to use existing models to represent static and dynamic behavior to an acceptable level of accuracy, but only in terms of terminal characteristics. This accuracy is generally measured with respect to either the original circuit behavior or specifications given in a data sheet for a particular part.

Macromodels typically require fewer components or fewer equations than the circuit level description, since no attempt is made to represent the internal phenomena in detail. This leads to faster computational times, smaller storage requirements, and the ability to simulate larger circuits.

3.1 Features of Macromodeling

As described in Chapter 2, macromodeling is a common method of creating simpler, more computer-efficient models for simulation [1]. There are several methods of macromodeling, some of which are described below.

Circuit simplification

One technique of creating a macromodel is to begin with the original circuit and perform some form of simplification (i.e., the removal of characteristics that are not necessary for the model). One form of simplification is to remove circuit elements from the original circuit one at a time, replacing them with open or short circuits as appropriate. This type of simplification is based on sensitivity estimates of the terminal characteristics of interest. As long as removing an element doesn't cause a significant change in the terminal characteristics that the designer needs in the final macromodel, it is acceptable to do so. One of the realities of this technique when applied to analog circuits is that in order to maintain a high level of accuracy, a large reduction in circuit components is not typically possible [2].

The next generation of this technique is to replace the element with a simpler model instead of an open or short circuit. An example of this would be replacing the active load of a differential amplifier with load resistors. Another example would be to use a Shichman-Hodges MOS model for the input devices of the MOS differential amplifier rather than the more complex model actually used to simulate the circuit such as BSIM, Level 2, or Level 3 [3-5, 7]. While the number of equations remains the same, the complexity of the models to be evaluated can be significantly reduced.

Circuit build-up

Another macromodeling method that is commonly used is referred to as the *circuit build-up* technique [8,9]. This type of macromodel produces a circuit configuration consisting of ideal elements. These elements are intended to meet the terminal characteristics or data book specifications without necessarily resembling a portion of the actual circuit configuration. This modeling process is very similar to circuit design except that in macromodeling one starts with ideal behavior and proceeds towards non-ideal whereas circuit design works in the opposite direction. Nevertheless, the thought processes are similar, thus this approach is the most widely used.

The circuit build-up process can be somewhat lengthy, since each proposed model configuration must be manually validated against the data book specifications or the performance of the actual circuit. By using an optimization scheme, this validation can be automated, which reduces the chance that some of the tedious validation steps may be neglected.

Symbolic macromodeling

A third technique that has shown some merit is macromodel generation from symbolic analysis [2, 10-13]. The symbolic macromodel is found by symbolic manipulation of the presolved original circuit. The manipulations traditionally consist of matrix formulation and ordering, Gaussian elimination, and forward substitution. These macromodels can be represented by controlled sources and basic circuit components, but they bear no resemblance to the original circuit.

Combining techniques

The most effective technique for macromodeling is combining the circuit simplification, build-up, and symbolic techniques described above. By applying the appropriate method to a given situation, a macromodel can be created more efficiently.

Figure 3-1 [8] shows the Boyle macromodel introduced in the first chapter, which is an example of a model derived by a combination of circuit simplification and circuit build-up. The input stage was derived using simplification, whereas the middle and output stage were derived using the build-up method.

Two characteristics common to macromodels are:

1. They are typically based on the structure or topology of the original circuit that they were intended to model.

2. They are implemented by using existing models for circuit level primitives (e.g., diodes, transistors, resistors, and capacitors) and controlled sources as predefined blocks. While this correlation may at times be less pronounced, this is typically the case only when the macromodel is intended to represent a class of circuits, such as op amps or comparators.

These characteristics of macromodels are important in that they distinguish macromodels from behavioral models. While

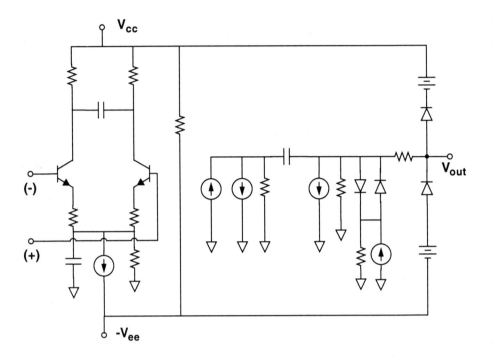

Figure 3-1. Diagram of the Boyle op amp macromodel (©1973 IEEE).

behavioral models may, in some cases, be "inspired" from the structural details of a circuit, the resulting model is implemented as a set of equations or other non-circuit-like functions.

One advantage of writing a macromodel is that it incorporates previously written code. Another advantage is that macromodels typically simulate faster than a primitive-level representation of the same circuit. The factor of improvement in simulation speed is typically between 5 and 10 [1,8]. However, a macromodel typically simulates slower than a corresponding behavioral model of the same circuit.

A primary disadvantage of macromodels is that they depend on the availability of underlying models. If one desires to model a switching power supply control IC, but building block models for comparators, oscillators, and drive stages are not available, then the task becomes prohibitively complex. This dependence on available models leads to a further disadvantage of using macromodels—the conceptual distance they can place between the intended overall model and its implementation with available primitives.

Another disadvantage is that, by being constrained to using only existing models, some macromodels entail a lack of computer efficiency. This is because the number of circuit nodes for the macromodel remains constant for all simulations. For example, the collector resistances in the first stage of a macromodeled op amp are provided as resistors between the V_{cc} supply and the collector terminals of their respective transistors (as shown in Figure 3-2). If these resistances were equal to zero, then nodes **Vcc**, **col1**, and **col2** would still remain in the model for the simulator to include in the solution matrix. With an AHDL, it is possible to write a model that will check whether either or both resistances are zero and then collapse nodes **col1** and/or **col2** to **Vcc** for improved efficiency.

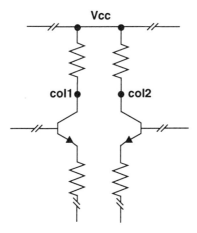

Figure 3-2. Collector resistances in the input stage of an op amp.

3.2 Example—Simultaneous Differential Equations

An example helps to illustrate many of the shortcomings that macromodeling can present. Consider the three simultaneous differential equations listed below. While these equations are presented without justification as to what they might represent, they serve to illustrate the ease with which an AHDL addresses modeling problems.

Equation	Initial conditions
$\dot{x} = y(1-z)$	$x = 18$
$\dot{y} = x(15+z) - 2\dfrac{y}{t}$	$y = 0$
$\ddot{z} = z - 2\dfrac{\dot{z}}{t} - 2(x^2 + y^2)$	$z = 7, \dot{z} = 0$

SPICE implementation

Figure 3-3 shows a block diagram of how these equations could be implemented using a SPICE macromodel. The SPICE netlist of the macromodel depicted in Figure 3-3 is shown below [14]. Note that a macromodel is merely an input file to the simulator—there is very little in the model code to indicate the how this macromodel implements the given set of differential equations.

```
EXAMPLE DIFFERENTIAL EQUATION SIMULATION
.OPTIONS ACCT NOPAGE
*
.SUBCKT INT 1 2
*CONNECTIONS: IN OUT/IC
*INTEGRATOR SUBCIRCUIT MODEL
*V(2)/V(1) = 1/S
GINT 0 2 1 0 1UMHO
CINT 2 0 1UF
*RINT PROVIDES A DC PATH TO GROUND
RINT 2 0 1000MEG
.ENDS INT
*
```

Macromodeling

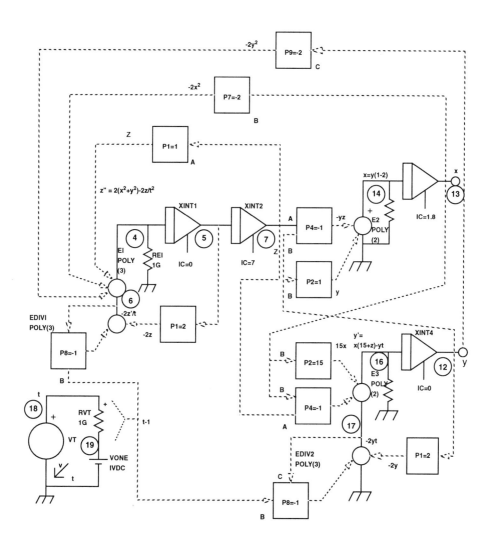

Figure 3-3. Block diagram of simultaneous differential equations implemented in SPICE (©1992 IEEE).

X1 4 5 INT
X2 5 7 INT
X3 14 13 INT
X4 16 12 INT
RE1 4 0 1000MEG

```
* A B C P0 P1 P2 P3 P4 P5 P6 P7 P8 P9
E1 4 6 POLY(3) 7 0 13 0 12 0 0.0 1 0.0 0.0 0.0 0.0 0.0 -2 0.0 -2
*V(4,6)=V(7)-2*V(13)-2*V(12)*V(12)
*
* A B C P0 P1 P2 P3 P4 P5 P6 P7 P8 P9
EDIV1 6 0 POLY(3) 5 0 18 19 6 0 0.0 -2 0.0 0.0 0.0 0.0 0.0 0.0 -1
*
RE2 14 0 1000MEG
*
E2 14 0 POLY(2) 7 0 12 0 0.0 0.0 1.0 0.0 -1.0
*
RE3 16 0 1000MEG
*
E3 16 17 POLY(2) 7 0 13 0 0.0 0.0 15 0.0 1.0
*
EDIV2 17 0 POLY(3) 12 0 18 19 17 0 0.0 -2 0.0 0.0 0.0 0.0 0.0 0.0 -1
*
VT 18 0 PWL(0 0 10 10)
RVT 18 19 1000MEG
VONE 19 0 DC 1V
.IC V(5)=0 V(7)=7 V(13)=1.8 V(12)=0
.TRAN 0.02SEC 2SEC 0.0 0.02SEC UIC
.PRINT TRAN V(13) V(12) V(7)
.PLOT TRAN V(13) V(12) V(7)
.END
```

Behavioral implementation in MAST

Although the MAST language is quite capable of using existing models to duplicate the implementation shown above, the three differential equations can also be modeled much more simply, as shown below. Note that, because it is a modeling language, MAST allows the expression of such equations in a much more direct manner. This flexibility eliminates the need to configure existing models, such as electrical elements, into an application for which they were never intended.

This example illustrates the difference between behavioral models and macromodels. Macromodels force the modeler into using a structural representation—even when one is not necessary or particularly desirable.

```
#Model for solving simultaneous nonlinear differential equations.
# Declare system variables
var nu  x, y, z, zprime
#Definition of equations
x:          d_by_dt(x) = y*(1-z)
y:          d_by_dt(time*y) = time*x*(15+z) - 2*y
z:          d_by_dt(time*zprime) = time*z - 2*zprime - 2*time*(x**2 + y**2)
zprime:  d_by_dt(z) = zprime
#Set initial conditions
control_section {
    initial_condition(x,1.8)
    initial_condition(y,0)
    initial_condition(z,7)
    initial_condition(zprime,0)
    }
```

Should these differential equations actually represent a physical system, it is a trivial matter (as shown in Chapter 6) to create a new simulation primitive out of this model using an AHDL. It would then become available for use in other, more complex, macromodels.

Conclusion

An AHDL still allows the use of netlists to create macromodels using known techniques. However, the realm of possibility expands considerably when the availability of underlying models is no longer a constraining factor, and when the number of underlying models has significant potential for growth.

3.3 References

1. *Linear Circuits Operational Amplifier Macromodels Data Manual*, Texas Instruments, 1990.
2. H. Y. Hsieh, N. B. Rabbat, and A. E. Ruehli, "Macromodeling and Macrosimulation Techniques," *Proc. 1978 IEEE Int. Symp. Circuits Syst.*, pp. 336-339, Apr. 1978.

3. B. J. Sheu, "MOS Transistor Modeling and Characterization for Circuit Simulation," Memorandum No. UCB/ERL M85/85, Electronics Research Laboratory, University of California, Berkeley, 1985.

4. B. J. Sheu, D. L. Scharfetter, P. K. Ko, and M. C. Jeng, "BSIM: Berkeley Short-Channel IGFET Model for MOS Transistors", *IEEE Journal for Solid-State Circuits*, Vol. SC-22, No.4, Aug. 1987.

5. B. J. Sheu, D. L. Scharfetter, and H. C. Poon, "Compact ShortChannel IGFET Model", *Memorandum No. UCB/ERL M84/20*, Electronics Research Laboratory, University of California, Berkeley, 1984.

6. A. Vlademerescu and S. Liu, "The Simulation of MOS Circuits Using SPICE2", *Electronics Research Laboratory*, No. UCB/ERL M80/7, 1980.

7. L. M. Dang, "A Simple Current Model for Short Channel IGFET and Its Application to Circuit Simulation," *IEEE Journal of Solid-State Circuits*, p. 14, 1979.

8. G. R. Boyle, B. M. Cohn, D. O. Pederson, and J. E. Solomon, "Macromodeling of integrated circuit operational amplifiers," *IEEE J. Solid-State Circuits*, Vol. SC-9, No. 6, pp. 353-363, Dec. 1974.

9. B. Epler, "SPICE2 application notes for dependent sources," *IEEE Circuits and Devices Magazine*, Simulation and Modeling Column, pp. 36-44, Sept. 1987.

10. M. H. Heydemann, "Functional macromodeling of electrical circuits," *Proc. 1978 IEEE Int. Symp. Circuits Syst.*, pp. 532-535, Apr. 1978.

11. N. B. Rabbat, A. L. Sangiovanni-Vincentelli, and H. Y. Hsieh, "A multilevel Newton algorithm with macromodeling and latency for the analysis of large scale nonlinear circuits in the time domain," *IEEE Trans. Circuits Syst.*, Vol. CAS-26, No. 9, pp. 733-740, Sept. 1979.

12. G. Gielen and W. M. C. Sansen, *Symbolic Analysis for Automated Design of Analog Integrated Circuits*, Kluwer Academic Publishers, 1991.

13. Q. Yu and C. Sechen, "Approximate Symbolic Analysis of Large Analog Integrated Circuits," *IEEE Int. Conf. on Computer- Aided Design (ICCAD)*, Nov. 6-10, Santa Clara, CA, 1994.

14. D. B. Herbert, "Simulating Differential Equations with SPICE2," *IEEE Circuits & Devices*, Vol. 8, No. 2, pp. 9-13, Mar. 1992.

CHAPTER 4

Top-Down Design

4.1 Introduction

The initial phase of a large project is usually characterized by a daunting number of details and preliminary block diagrams. In confronting the conceptual distance between the block diagram level and the details of implementation, designers over the years have developed and refined an approach called *top-down design*.

What is top-down design? Basically, it is the process of evolving a design from initial concept to the final details in a systematic and analytic fashion. Typically, this technique takes advantage of common design and verification tools such as computer simulation. Although it is not the intent of this book to provide instruction on top-down design, the capabilities of an AHDL can make top-down design easier, both in conceptualization and implementation.

Top-down design helps manage the complexity of projects that have more variables and levels of detail than one person can reasonably be expected to handle. For example, the maximum complexity most designers are comfortable with is that which can be shown in a block diagram on a standard size piece of paper. Any more information than that is generally too much detail. By applying top-down techniques, a person or a group can work on large projects with maximum efficiency and flexibility. While an AHDL may not be a requirement for top-down design, the advantages of an AHDL may eventually make it a requirement for managing design complexity.

Figure 4-1 shows a typical representation of the top-down approach. Beginning at a sufficiently high level of abstraction, it should be possible to represent the system in a block diagram on a single piece of paper. By looking at sections of a design at various levels of abstraction, a designer can observe and manage the interactions between sections more easily. Each block can then be evaluated as to whether it can be designed at a lower level of abstraction. This evaluation continues until the abstraction level of each block has been reduced to a size that is suitable for implementation. At this point, purchasing or prototyping can be considered.

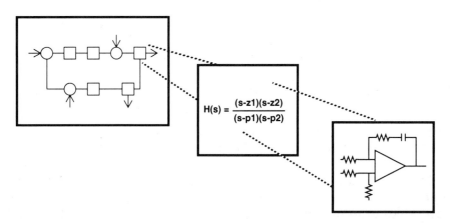

Figure 4-1.   Levels of abstraction in a top-down design.

Linking top-down design methodology with simulation is a powerful combination that allows finding problems at each stage early in the cycle when they're easier to correct. The alternative can be expensive and inefficient retrofits. Further, the analysis techniques are available at each abstraction level. For example, a Laplace transform (the middle block in Figure 4-1) is much easier to use with a control system block diagram than with transistor circuitry.

With an AHDL, it is possible to create models that span several levels of abstraction. At the higher levels, accurate behavioral models used by system design groups can significantly increase simulation speed. At the lower levels, traditional "primitive" models can accept characterization data from electrical and mechanical design groups.

There are two basic skills that come with practice in using top-down design:

1. Developing a knack for determining where to break the design into sub-units. This is more difficult with analog design than it is with digital design. Digital circuitry is designed so that each output signal is relatively unaffected by the other devices on that signal line. This is typically not the case for analog circuitry where outputs and inputs interact with one another—the partitioning is usually performed based on function (e.g., filters, amplifiers, etc.).

2. Knowing what the appropriate levels of abstraction are. General intuition is often a good indicator as to whether a level has been skipped and whether too much is trying to be accomplished at once.

4.2 An Electro-mechanical Motion Control System

Figure 4-2 shows a top-level block diagram of a shaft angle controller being designed by a control system engineer. This could be part of any servo-type system, such as an engine control actuator or a radar antenna positioner. The individual transfer function (Laplace transform) blocks represent the various dynamic elements of this system.

The main purpose of this example is to highlight the biggest advantage of the top-down approach: that higher level design information can be passed down to other groups (such as manufacturing, test, and reliability) before a prototype or "breadboard" is even built. Timely feedback from these groups can reduce surprises at later stages of product development.

Simulation with AHDL models allows analysis and verification of an entire system, from the initial design concept (block diagram) to the actual electrical and mechanical implementation.

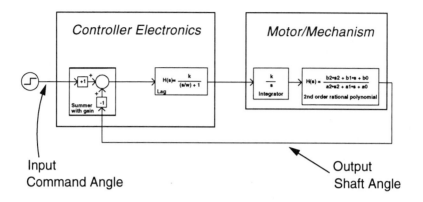

Figure 4-2. Top-level design of a shaft angle controller.

This capability is essential for the design of complex motion control systems, such as those found in automobiles, aircraft, or space vehicles.

Simulating the top-level design

Figure 4-2 shows four top-level design blocks contained within two different modeling domains—electrical and mechanical. With a hardware description language, the blocks at any given level can be made available as library models (as has been done with MAST for the Saber simulator). The ability to provide linear and nonlinear transfer functions, including rational polynomials (Laplace functions), integrators, limiters, and hysteresis demonstrates the utility of an AHDL for top-down simulation.

At the left of Figure 4-2, an input voltage (possibly from a transducer) represents the rotation angle to be applied to the controlling electronics. This input signal is then compared with the measured angle of the output shaft, which produces an error signal. The error signal is filtered by a lag compensator and is then used to drive the motor and respond to mechanical load dynamics.

At this level, a designer can use simulation to adjust the compensator gain and bandwidth, specify the maximum inertia or mechanical load, or specify the required torque capability of the motor. With the appropriate parameters in place for the block models, system performance can be "tuned" to achieve the desired step response resulting from a transient analysis (Figure 4-3). This response has a rapid rise, 35% overshoot, and a short settling time.

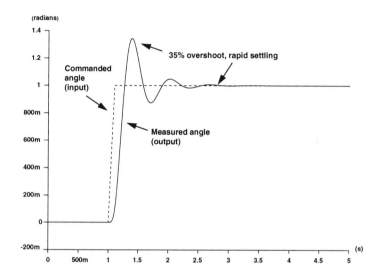

Figure 4-3. System response of the top-level design.

Moving down—refining the design

Once the system design group has obtained the desired response from a top-level simulation, the corresponding subsystem specifications may be given to the mechanical and electrical design groups. These groups can work independently to provide the details of the final implementation.

Figure 4-4 shows how the finished mechanical design might look. It can be inserted into the original system model to replace the *Motor/Mechanism* block in Figure 4-2. This design now consists of true mechanical models and can be used for performance verification.

As a further enhancement to this example, assume that mechanical "parasitics" in the Gearbox are a key area of concern for this portion of the design. It is possible to include effects such as friction, shaft flexibility, and backlash in the Gearbox model. Each of these effects produces a unique and troublesome "signature" on the closed loop response, as shown in Figure 4-5.

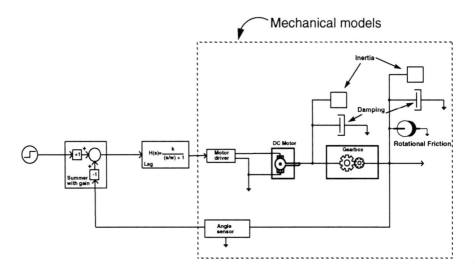

Figure 4-4. Inserting mechanical models into the original top-level design.

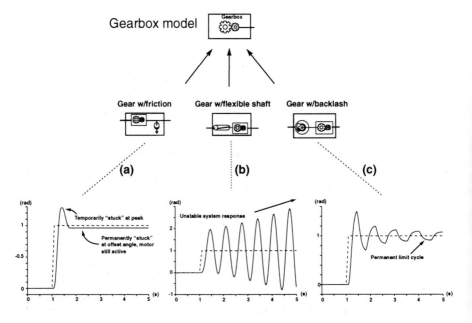

Figure 4-5. Three undesirable mechanical effects in the Gearbox.

Figure 4-5a shows a friction effect, which causes the response to stick each time the shaft comes to rest. This finally causes the shaft to remain stuck at a permanent offset angle. Note also that the motor continues to exert a torque to reduce this offset, but it is unable to overcome this static friction. Even though the offset may be tolerable as far as satisfying the systems positioning requirements, the waste of energy and wear on the motor are likely to prove unacceptable.

Likewise, Figure 4-5b shows the effect of the flexible shaft, which causes even less acceptable performance. Because of this flexible bending effect in the shaft, the error compensator can never catch up to change in shaft position. This causes the control loop to become completely unstable.

Finally, Figure 4-5c shows the effect of mechanical backlash in the gearbox, which tends to decouple the input and output gear shafts for small angle variations. This prevents the loop from settling at the commanded angle once it is near the desired value. A permanent limit cycle or "hunting" behavior is observed, which results in unacceptable wear and inefficiency. This particular effect is examined below, with a proposed solution that is conceived at one level and implemented at another.

Investigating a solution for backlash

Because it is still early in the design process, the system designer still has several options to fix any of these problems. For example, the specification can be tightened on the gear backlash (Figure 4-5c) by precision machining of the teeth, or by adding an anti-backlash spring. Unfortunately, both these solutions would increase the manufacturing costs beyond competitive limits. However, careful review of the system performance specifications shows that a small, steady-state offset angle is acceptable, as long as there is no "hunting" or energy dissipation.

This leads to the consideration of whether something might be added to the *Controller Electronics* block to "capture" inputs in this acceptable range and alleviate the problem before it propagates to the *Motor/Mechanism* block.

One such possibility is the insertion of a deadband model in the controller block. This can be written as a behavioral model

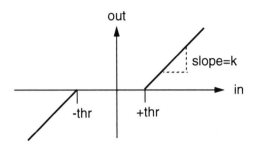

Figure 4-6. Deadband behavior.

that defines three piecewise linear regions of behavior (this is covered more thoroughly in Chapter 8; see Appendix B for a source listing of the template for this model).

As shown in Figure 4-6, the deadband model defines upper and lower input limits that determine the one of three possible outputs:

- If the input is less than the negative threshold (**-thr**), the output is **k*(in + thr)**.
- If the input value is between the deadband limits, (**-thr** and **+thr**), the output is zero.
- If the input (**in**) is greater than the positive threshold (**thr**), the output (**out**) is **k*(in - thr)**.

The values for the gain (**k**) and the symmetrical threshold (**thr**) are user-specifiable parameters of the deadband model. The gain can be any value (including negative values) and has been given a default value of 1. The threshold value must be nonzero, and it too has default value of 1.

The input and output connections are modeled as unitless variables instead of currents and voltages. This avoids the need to match units, provide reference nodes, or provide return paths to ground.

Figure 4-7a shows the insertion of a behavioral deadband block back into the original system design. Notice in Figure 4-7b that the step response of the output shaft angle still overshoots the commanded angle, but then it settles—first at a large offset, then at a smaller offset where it remains. This offset is within the acceptable performance range, and the motor is off.

The major point to observe is that the behavioral model for the deadband controller *required no actual electronics to be built or even designed!* Because the specification for this deadband function can be added so easily (i.e., the electronic design is not yet necessary), there has been only minimal cost incurred and no hardware to scrap.

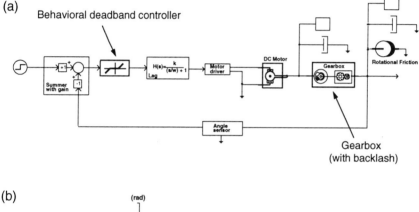

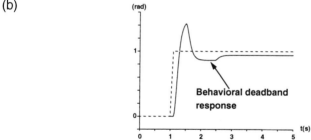

Figure 4-7. Inserting a behavioral deadband block to correct backlash.

Full system implementation

Once the deadband block for the backlash problem has been verified, the electrical and mechanical implementations of the overall servo system can now be modeled and simulated. Figure 4-8a shows this implementation, replacing the control blocks with the electrical circuits that perform those functions (see Appendix

B for a netlist of this design). A voltage source represents the commanded angle, and a differencing network produces the error signal. This is applied to a zener bridge network, which provides a simple electrical implementation of the deadband controller. When the error signal is large, the zeners provide a current path to the next gain stage. Small error signals are blocked, so no drive is applied to the motor. The lag compensator is a simple RC filter.

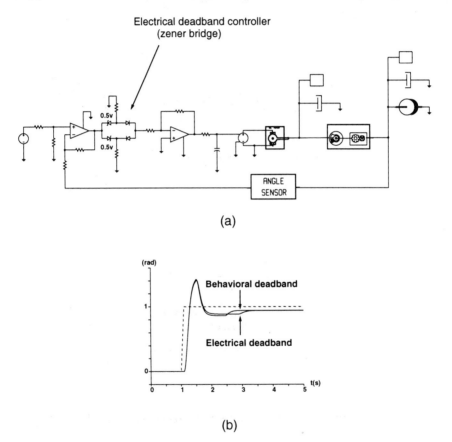

Figure 4-8. (a) Inserting electrical models into the overall design and (b) comparison of the behavioral and electrical results.

Figure 4-8b shows the step response for this full circuit/mechanism implementation. Note that the final correction occurs after a slightly longer delay due to non-ideal electrical effects that were not present in the ideal control blocks. The final offset angle is nearly identical, however. Although an ideal motor driver is used here, a more up-to-date design using power devices (such as IGBTs and power MOSFETS) could have been created from existing model libraries.

Conclusions

With some confidence in this implementation, the system level designer can now pass along this preliminary information to other design and support organizations, such as manufacturing, reliability, and test. This early exchange of data provides important interaction between organizational groups. For instance, manufacturing may point out that the 0.5V zener diodes are unavailable on their Standard Parts list, so a suitable alternate value must be selected. Similarly, the test group may warn of the need for a velocity test point in the sensor section so that they can perform a required step in the test procedure.

All of this participation between design and verification groups provides opportunities for significant cost savings by avoiding potentially major problems. For instance, the 0.5V zener diodes could be replaced with ones rated at 2.5V simply by making minor adjustments to the adjacent gain stages.

This example has demonstrated one instance of using simulation and modeling as design and analysis tools throughout the life cycle of a system project. These phases range from the initial concept through specifications, design, implementation, and testing. Top-down design followed by bottom-up verification can uncover potential problems early in the design process. Evaluating the trade-offs across interfaces (i.e., between model technologies) helps find these problems during simulation and allows changes to be made with minimal cost and schedule impact.

4.3   References

1. *Top-Down Design:Motion Control Applications*, Publication MA-0149, Analogy, Inc., June 1992.

Section 2

Model Implementation
Chapters 5-12

This section begins to convey the implementation details of the MAST AHDL. The chapters contained in this section show how to use the governing equations of several commonly used models, along with equations that are readily available from well-known textbooks and papers. This information is provided in both tutorial and reference fashion, serving as an introduction to the basics of the MAST AHDL.

CHAPTER 5

Connecting Models to the MAST AHDL

The intent of this chapter is to sharpen the context of implementing a model in an AHDL. The goal is to proceed from the "thought processes" and equations of the modeler (introduced in Chapter 2) to a simulation model for an end user. In MAST, the principal mechanism for making this connection is the template.

5.1 What is a Template?

A template is a model written in the MAST modeling language for use by the Saber simulator; it communicates the mathematical and/or logical functions performed by a design element to the simulator. A template is contained in a text (ASCII) file that is given the extension .sin, which stands for *Saber in*put.

Although it must be clearly understood that MAST is not a general programming language, a template does have some organizational requirements that loosely parallel those of a computer program. The most prominent of these is the organization of a template into sections, which provide consistent and meaningful locations for the implementation of the thought processes depicted in Figure 2-2. Other similarities include classifying variables according to various data types, defining a variable before using it (including its data type), and observing specific meanings of certain characters and words.

An overall procedure for implementing these thought processes is described more specifically in Section 5.2.

A few definitions

Although it is not the intention of this chapter to provide rigorous instruction on all aspects of writing a template, defining a few terms here will prove useful later.

template—as described above, a model written in the MAST modeling language.

declaration—the identification and definition of a template variable. Some variables need to be declared before they can be used in the template; for others, declaration is optional.

section—any of several template partitions that can be used when writing a template to identify where certain calculations, declarations, comparisons, and assignments are being performed.

reserved word—a set of characters that has special meaning in the MAST language (such as **template** or **number**). When writing a template, these words should be used only for their assigned purposes. For example, do not use **number** as the name of a variable in a template.

parameter—a type of template variable that assumes a constant value throughout simulation.

argument—a parameter for which the template user can specify a value.

netlist—a listing of a particular network of models (templates) for a specific simulation.

node—point at which two or more templates are connected in a netlist.

instance—the occurrence of a template in a circuit (netlist). This includes the connections to other templates and any user-specified argument values.

reference designator—a unique name for a template instance in a design. Typically, this is an alphanumeric identifier that is separated from the template name by a period (such as **r1** in the following example).

A quick example

The following example illustrates several of the terms listed above. It shows how a template user would create a line in a netlist for one instance of one template (named **resistor**).

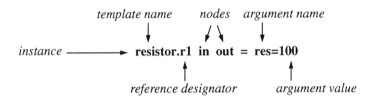

The simulator would read this line in the *netlist* as an *instance* of the *template* named **resistor**. The simulator would assign a *reference designator* of **r1** to **resistor**, connect it to system nodes **in** and **out**, and use 100 for the value of the **resistor** *argument* named **res**.

5.2   Template Organization and the Modeling Process

Basic template organization

The conventional method of organizing a MAST template is shown below. This conforms to basic structural requirements that are part of any language. Although there is some degree of flexibility to this that will be exercised in the examples throughout the rest of the book, this organization lends itself well to accomplishing the goal of this chapter.

> **Standard Template Layout**
>
> **unit definition (automatically included by Saber)**
> **pin definition (automatically included by Saber)**
> **header**
> **header declarations**
> **local declarations**
> **parameters section**
> **netlist statements**
> **when statements**
> **values section**
> **control section**
> **equations section**

NOTE

Note that the shaded entries (unit and pin definitions) are shown here only for their contribution to the modeling process. In practice, they are defined in a file used by all templates—they are almost never required when writing a template (i.e., for all practical purposes a template starts with the header).

In particular, Appendix A contains a more thorough listing of MAST reference information such as syntax, parameter types, functions, and foreign routines.

Dependent and independent variables

As explained in Chapter 2, the ultimate task of a continuous-time simulator is to solve a system matrix that represents a system of simultaneous equations. Each such equation is formulated so that a dependent variable is expressed in terms of independent variables. A model written in the MAST AHDL is simulated using a *modified nodal analysis* technique. In nodal analysis, there is a through and an across variable associated with each node in the design. Through variables are directly added to (or subtracted from) the system matrix, and the simulator then solves for across variables. Therefore, when this is the case, the through variables are dependent variables and across variables are independent variables.

Some models, however, implement the through variable as an independent variable. For example, the model of an ideal voltage source defines the through variable (current) as whatever it needs to be to maintain the specified across variable (voltage). In such a case, the through variable is an independent variable—it is *not* defined in terms of the across variable. This requires adding an equation for the through variable to the system of equations.

In general, most through variables are dependent variables, and all across variables are independent variables.

Philosophy of writing a simulation model

When writing a template, the writer (modeler) must accommodate the following requirements:

- The characteristics of the simulator
- What the modeler intends to provide for simulation
- What the user of the model expects to obtain by using the model and the simulator

The simulation process described above is predicated on the notion that the simulator takes an initial guess at the solution of the system of equations based on the information from the models. This guess is then refined at each iteration of the given analysis. The simulator will make these refinements based on its internal numerical algorithms. This may result in unusual or irrelevant calculations unless the simulation model (template) specifies otherwise. In other words, the simulator has no "knowledge" of the areas of interest (such as a discontinuity or a region of an exponential function) unless the template has been explicitly written to limit simulation to those areas.

In addition, the writer should account for what kind of information the user can provide as argument values. For example, if the user is most likely to have capacitance values, but the model uses charge in its characteristic equations, then the writer needs to permit the user to specify capacitance and then convert those values to charge inside the template.

Expressing the basic model

Figure 2-2 gives a descriptive overview of how to go from formulating ideas for a model to expressing those ideas in terms of model implementation. Figure 5-1 extends this overview by correlating implementation requirements to actual template sections.

Figure 5-1a depicts the process of model development in terms of the following principal elements:

- Usage
- Dependent variables
- Independent variables
- Parameters
- Simulation variables and "environmental" effects

Basically, the process of writing a model in an AHDL consists of two fundamental steps:

1. Adequately defining the model in terms of these elements, which is a prerequisite to writing the template.
2. Putting the proper expressions in the corresponding template sections (Figure 5-1b), which shows a condensed representation of the organizational listing given on page 54. Section 5.3 provides a more elaborate explanation of each section and what it accomplishes.

The correlations of Figure 5-1 are described briefly below, using a hypothetical template named **resistor** (introduced on page 53).

Connecting Models to the MAST AHDL

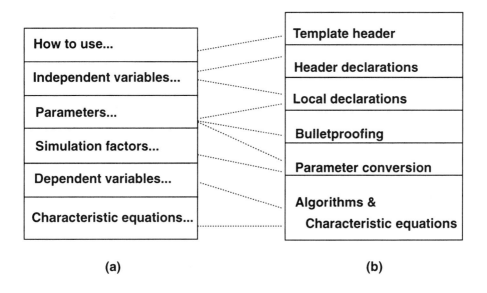

Figure 5-1. Correlating a model to a MAST template.

How to use—The template header defines the template name, its connection points (which automatically include unit and pin definitions), and a list of argument names. Each of these items is a part of netlist usage. Here, using the netlist entry for the **resistor** template shown in Section 5.1, the header would be written to establish a correspondence to this netlist entry:

resistor.r1 in out = res=100 ⬅——————— *netlist entry*

template resistor p m = res ⬅——————— *template header*

The netlist entry designates this instance of **resistor** as **r1**, connects pin **p** of the template to node **in**, connects pin **m** to node **out**, and specifies the value of **res** as 100.

Independent variables—Given that the characteristic equation of the resistor model is some expression of Ohm's law ($I = V/R$), the independent variable is identified as the electrical across variable, voltage.

Here, a variable (called **vres**, for example) could be declared and computed as:

val v vres
vres = v(p) - v(m)

where **v(p)** and **v(m)** are the values of the across variable for this template. *These variables are calculated at each iteration of the simulation.*

Parameters—The constant for this model is the value of resistance. Most often, this value is provided as an argument for the user to specify in a netlist, in which case this section would be empty. Typically, however, a condition check for a nonzero value is done. This could provide a parameter that assumed the argument value if the value were nonzero, or that assumed some writer-defined value (say 1 mΩ) if the argument value were zero. The parameter would then be used in the characteristic equation instead of the argument. *These variables are calculated once—when the netlist is read—and remain constant throughout a given simulation.*

Simulation factors—A typical occurrence of this is temperature. MAST uses a variable called **temp** (the temperature at which the simulation is performed) that can be used to calculate a new value of resistance based on equations for temperature coefficients. This calculation would be performed on the argument value, the results of which could again be assigned to a parameter to be used in the characteristic equation instead of the argument. *These variables are calculated once—when the netlist is read—and remain constant throughout a given simulation.*

Dependent variables, characteristic equation—Given that the characteristic equation of the resistor model is some expression of Ohm's law (I = V/R), the dependent variable must be identified as the electrical through variable, current. *These variables are calculated at each iteration of the simulation.*

Here, the current, **i**, through the resistor (from pin **p** to pin **m**) is defined in the characteristic equation as:

i(p->m) += vres/res

5.3 Template Sections

Figure 5-2 shows the generalized form of a template, including some of the more commonly used constructs such as syntax, declarations, and sections. These are discussed in some detail below; Appendix A contains a more thorough listing of MAST reference information such as syntax, parameter types, functions, and foreign routines. Note that, except for the values and equations sections, *section variables are calculated once—when the netlist is read—and remain constant throughout a given simulation.*

Comments

The MAST language parser ignores tabs, spaces, blank lines, the pound sign (#), and any characters following a pound sign to the end of that line. Thus, any text following a pound sign (which can appear anywhere on a line) can be used as a comment.

Header

The header declares the name of the template (**name**), the names of the connection points (**pin1, pin2**), and the names of the arguments (**arg1, arg2**). The names of the arguments are separated from the names of the connection points by an equals sign (=) and from each other by commas.

The template header is required for a model primitive (i.e., if it is to be used in a netlist) and appears in the following general form:

>   **template** *templatename connectionpoints* = *arguments*

where **template** is a reserved word that identifies the contents of this file as a template, *templatename* is the "official" name of the template for use in a circuit description (i.e., a netlist), *connectionpoints* are the names of connections to the template, and *arguments* are the names of user-specifiable parameters. The actual names given to *templatename*, *connectionpoints*, and *arguments* are all selected by the author of the template.

Note that a template may consist of only a netlist or a netlist plus equations or other MAST statements, in which case the template would not require a header (see Chapter 6).

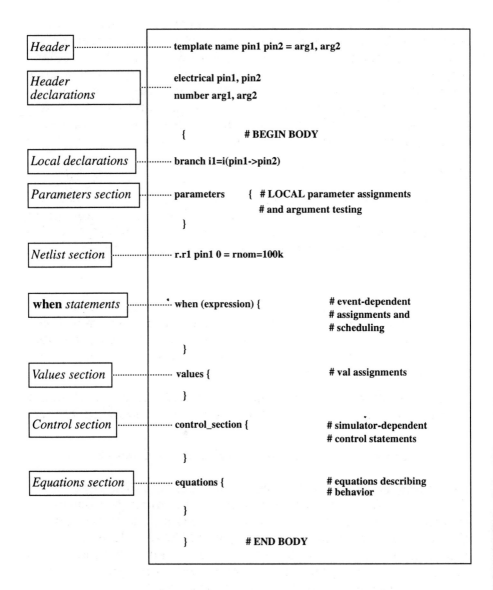

Figure 5-2. Sections of a MAST template.

Header declarations

Most variables within a template must be explicitly identified so that they can be recognized by the simulator throughout the rest of the template. This is called *declaring* a variable.

The header declarations section declares the following types of variables:

- connection points (appearing in the header)
- arguments (appearing in the header)
- external variables (to be passed in from hierarchy)
- exported variables (to be passed out to hierarchy)

Template body

The template body is a mandatory partition that separates the header and header declarations from the rest of the template. The body must explicitly begin with an opening brace ({) and terminate with a closing brace (}). Each brace should be on its own separate line. The body contains all template statements, sections, and declarations, except for the header and header declarations.

Local declarations

The local declarations section declares the following types of variables:

- branches (see page 65)
- local parameters
- local **vals**
- local **vars**

where a **val** is a quantity that can be assigned a value in the values section and a **var** is a quantity that cannot be assigned a value by the template, but is considered an independent variable that the simulator solves for iteratively.

This section is not explicitly declared (i.e., it has no opening and closing braces), but local declarations must appear before any other sections in the body. Local declarations mean exactly what the name implies. The **vals**, **vars**, parameters, and internal pins (or branches) are local to this template and are not accessible outside the template except for monitoring.

Parameters section

Model parameters are typically coefficients of the governing equations of a model. By "parameterizing" the model based on these coefficients, the model is capable of predicting the behavior of different specific devices within the more general family of devices for which it was intended.

An *argument* is a parameter whose value can be specified by a user in a netlist. Other parameters are local—the template uses them as it does arguments, but their values are not specified in a netlist.

Statements in the parameters section cause the simulator to perform one-time calculations upon invocation during the setup phase (i.e., just before displaying the **saber** prompt). One of the main uses of this section is to check argument values to make sure they don't create calculation problems (such as dividing by zero or taking the square root of a negative real number). This is known as *bulletproofing* the template. Units conversion and assignment of values to local parameters are also done here.

Parameters used in the equations section (see below) are often derived from multiple parameters. One such case is parameter dependency on temperature. The temperature variation of model parameters is typically processed in this section.[1]

This section requires an identifying reserved word (**parameters**) along with opening and closing braces.

Netlist section

This section contains the netlist description of another template or templates used hierarchically (see Chapter 6). This section is not explicitly declared (i.e., it has no opening and closing braces), and netlist entries can be listed in any order.

---

[1] For models that contain self-heating effects, the temperature variation is performed in the values section since temperature is a quantity that the simulator computes similar to node voltages.

When statements

A **when** statement is used to schedule events for discrete time simulation (see Chapter 10). It is not actually a section, but a type of statement that processes events in digital and mixed analog-digital modeling. A when statement appears outside of sections, and it is not uncommon for a template to have more than one of them. Only **state** variables can be assigned values in a when statement.

A **when** statement is procedural, so it is executed in sequential order. A when statement is used with one of the following intrinsic, event-driven functions:

- schedule_event
- event_on
- handle
- deschedule

These functions allow the manipulation of an event queue that exists in the simulator.

Values section

The algebraic relationships for the dependent variables of the model are implemented in the values section. Values are assigned to these dependent variables which have been declared as **vals** in the local declarations. Many of these **vals** are elements of the differential equations that are implemented in the equations section (see below). The values section is a procedural section, so statements are executed in sequential order. *These variables are calculated at each iteration of the simulation.*

This section requires an identifying reserved word (**values**) along with opening and closing braces.

Control section

This section defines simulator-dependent implementation details, generally for nonlinear models. It is fundamentally a declarative section, although it can contain if-else statements whose conditions are based on constant parameter expressions.

Post-simulation statements specific to the model may also be defined in the control section. Appendix A contains more information on these types of statements.

This section requires an identifying reserved word (**control_section**) along with opening and closing braces.

Equations section

This section contains the terminal (connection point) equations of the model, which indicate how the characteristic equations of the model are related to one another. Relationships involving the through and across variables must be defined in this section. *These variables are calculated at each iteration of the simulation.*

The characteristic equations implemented in a template can be any combination of linear or nonlinear algebraic or differential equations. No integral expressions are allowed; however, by differentiating both sides of an integral equation, you can convert it into a differential equation. Some templates may require additional modifications to get the equations into a usable form.

This section requires an identifying reserved word (**equations**) along with opening and closing braces.

Simplified MAST—sections can be optional

With the exception of the control section, *declaring sections within the template body is optional*; it is most often done for clarity within large, complicated templates (such as the diode and MOSFET templates in Chapter 9). That is, sections help document the template by identifying segments where specific tasks are isolated. They can also be used to optimize simulation efficiency, although techniques for doing so are not covered in this book. Consequently, model implementations are provided in a straightforward manner and without justification of all details (i.e., templates may be provided with or without sections, with the focus on the modeling issue at hand rather than on the mechanics of providing template sections).

Templates written without sections most often omit the values, parameters, and equations sections. The expressions, declarations of **vals** and **vars**, parameter checking, and equations that

Connecting Models to the MAST AHDL **65**

appear in these sections are written the same as with sections. In addition, two MAST constructs—**branch** and **make**—are provided that allow more compact methods of declaring variables and writing equations.

> **branch**—A conservative model (such as an electrical model that observes KCL and KVL) provides a *branch* when the model can connect between two nodes in a system and model a path for the through variable (e.g., current) from one node to the other. For example, the model of an electrical resistor (having two pins) provides a single branch in a circuit—from **p** to **m**. Similarly, the model of a bipolar transistor (having three pins) provides three branches: from base to emitter, from base to collector, from collector to emitter. This concept of branch from electrical circuit theory can be generalized to any conservative model and used as a MAST construct. When writing a template for a conservative model, you can define the through and across variables for each branch as explained below.
>
> Specification of a branch consists of the following:
>
> 1. The reserved word **branch**.
> 2. The branch variable, which is either a through branch or an across branch.
> 3. The declaration of the through or across variable.
> 4. The names of the two pins defining the branch, which must match the pin names in the header declaration and must be of the same type (e.g., **electrical**). For a through variable, pin names are separated by the combination of a hyphen and a right angle bracket (->). For an across variable, pin names are separated by a comma.

For example, the branch current and branch voltage of a resistor might be declared as shown below. Each number refers to the list given above.

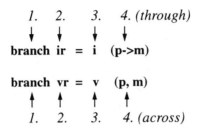

make—When the lefthand side of an equation is not a simple output, but an expression, it means the relation cannot be expressed directly. Statements such as this must be prefixed with the reserved word **make** so it will be recognized as a template equation. This kind of a statement is called a *constraint equation*. The word **make** is required anytime the output is an expression; it can also be used (although it is not required) if the lefthand side is an output. For example, the inductor template in Chapter 7 uses **make** because the lefthand side of the characteristic equation contains the through variable (current) in the expression for differentiation.

CHAPTER 6

Netlists—Using Hierarchy

This chapter provides a starting point for using some of the concepts and constructs described in preceding chapters. All remaining chapters in this book contain examples that serve the following main objectives:

- First, each example provides a walk-through of *how* MAST can accomplish a particular modeling task, which demonstrates the utility and power of an AHDL.

- More importantly, each example brings emphasis to one or more general modeling issues that need to be addressed.

The example in this chapter highlights the advantages of using hierarchy to create modules. It shows the details of putting together an active filter from existing models, how to make that filter usable as a single module, and how it compares to a behavioral model that provides the same functionality.

6.1 Modularity and Hierarchy

Chapter 4 described the concept of modeling at different levels of abstraction. One advantage of modeling with an AHDL is that it is possible to implement the appropriate level of abstraction and easily replace a given model with one of either higher or lower abstraction. This flexibility is achieved by employing modularity and hierarchy using some of the MAST constructs introduced in Chapter 5.

A template is the basic unit of modularity in the MAST language; it can provide the description of a single element in a design, a combination of several elements in a design, or the entire design itself. Central to the notion of modularity is the concept of the *netlist*, which is a list that represents how a network of template instances function together as a system. This list is provided as an input text file to the simulator. Each design element specified in a netlist calls a MAST template, which takes the user-specified argument values and plugs them into the equations, expressions, and assignments of the model.

Because it is a unit of modularity, a template can also be composed of further netlist entries, which introduces the concept of *hierarchy*. Those familiar with SPICE should realize that, although a MAST template is similar to a subcircuit (.SUBCKT), a template can also be extended to contain algebraic and differential equations directly—a profound distinction. Figure 6-1 depicts this relationship for three instances of a resistor template (specified with identifiers **r1**, **r2**, and **r3**), one instance of a capacitor template (specified with identifier **c1**) and one instance of an operational amplifier (specified with identifier **u1**). In MAST, the file containing the netlist of these template instances can also contain additional equations and statements that are not provided as part of any existing model (as indicated by the shaded area in the figure).

The use of modularity and hierarchy are good modeling practices. As in writing any software, it breaks a large problem into small, easily-identifiable pieces, which provides several advantages:

- Modules can be reused in other designs.
- One model can be replaced with another having a different level of abstraction (especially useful in top-down design).
- Debugging is made easier.

- Partitioning of design tasks can be done according to modules.

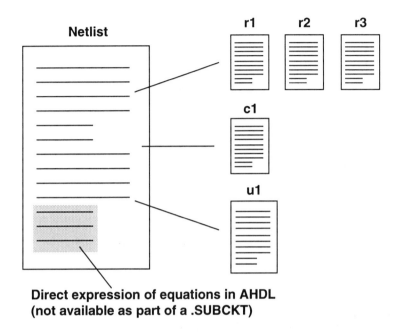

Figure 6-1. Typical hierarchical modeling.

In addition to simplifying implementation, hierarchical grouping of templates can provide increased computational efficiency. For example, the numerical methods used by the Saber simulator to solve systems of equations can take advantage of extra information in the specification of a hierarchical design. The internal matrix representation of the system of equations preserves and exploits the hierarchy. The efficiency benefits can be substantial, especially in large, highly-repetitive systems.

By contrast, some simulators do allow the hierarchical *specification* of a netlist, but then they "flatten" the netlist before simulation. This means that the internal matrix representation of the set of differential equations is the same as if the design were entered in a one-level netlist with no hierarchy.

6.2 Constructing a Module for an Active Filter

Consider an example of a simple low-pass filter. Figure 6-2 shows a block representation of the filter with a given low-frequency gain and a -3dB frequency (**fp**).

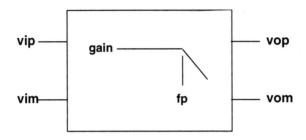

Figure 6-2. Block representation of a low-pass filter.

In the context of using the top-down design principles described in Chapter 4, a behavioral model (implemented as a transfer function) for this filter module might be appropriate. Such a model is listed in Appendix B. A designer could then use such a model to work at a higher level of abstraction when beginning on a circuit in which the filter would appear; it is much easier to specify the filter in terms of gain and frequency than to be constantly deriving appropriate component values.

Once the overall design is functioning satisfactorily at this level, the filter module can be implemented at a lower level of abstraction. Figure 6-3a shows how existing models can be used to construct the filter from its underlying components of an operational amplifier, three resistors, and a capacitor. Figure 6-3b shows a symbol for these components functioning together as a low-pass filter.

Netlist usage

The netlist that defines the network of Figure 6-3a is shown below. Note that there is a one-to-one correspondence between the elements in the filter module and the entries (lines) in the netlist.

Netlists—Using Hierarchy

(a)

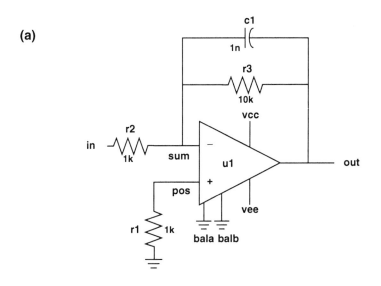

(b)

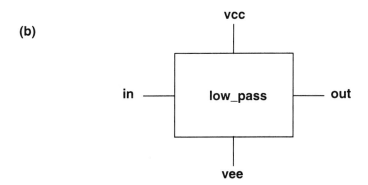

Figure 6-3. Building an active filter module (a) the primitive models to be netlisted (b) the hierarchical module.

```
r.r1 pos 0        = 1k
r.r2 in sum       = 1k
r.r3 out sum      = 10k
c.c1 out sum      = 1n
op42fz_2.u1 sum pos out 0 0 vee vcc
```

As described in the previous chapter, each line (netlist entry) defines an instance of a given model (i.e., its argument values and circuit connections) for use by the simulator and consists of the following parts:

- The name of the model.
- A period (.).
- The name of the instance of that model (sometimes called a *reference designator*).
- The names of the nodes to which this model instance is connected.
- An equals sign (=).[1]
- Values for the template arguments.[1] These can be either numerical values or expressions of other parameters that have numerical values assigned to them. These parameters can be arguments from other templates in the netlist.

Although schematic entry programs and automatic netlisters eliminate the need for creating netlists manually, there are some benefits to learning the syntax:

- Some concepts (such as argument expressions) are not easily discernible from a schematic entry program.
- Netlist statements may be used when writing templates. It is often easier to learn the syntax and write such a template by hand than to try to construct the hierarchy from a schematic entry program.

Creating a hierarchical template

The netlist for the filter module can be included in a template named **low_pass** (shown below), using the conventions and constructs for templates described in Chapter 5.

The ability to be able to pass expressions of arguments through the hierarchy is an important feature of an AHDL.

---

[1] Not present when user-specified parameter values are omitted, as in the case for op42z_2.ul.

Constructing a hierarchical template is a fairly straightforward process—each line of low_pass is explained below.

```
template low_pass in out vcc vee = gain, fo, rin
electrical in, out, vcc, vee
number    gain=1,            # low frequency gain
          fo=1k,              # corner frequency
          rin=1k              # input impedance
{
number    pi = 3.14159       # local declaration of a constant
# Netlist of filter module
r.r2 in sum  = rin
r.r3 out sum = rnom = rin * gain
r.r1 pos 0   = rnom(r.r2) * rnom(r.r3) / (rnom(r.r2)+rnom(r.r3))
c.c1 out sum = 1/(2*pi*fo*rnom(r.r3))
op42fz_2.u1  sum pos out 0 0 vee vcc
}
```

template low_pass in out vcc vee = gain, fo, rin

This is the template header. It begins with the MAST reserved word **template** followed by the name of the template (**low_pass**). The four connection points (**in, out, vcc, vee**) of the filter module are listed next, followed by the argument names. User-specified values for these arguments are passed to the underlying templates in the netlist.

electrical in, out, vcc, vee

This is the declaration of template connection points, which must match the connection point types of the underlying templates that provide connections of the module. Here, they are declared as electrical pins, which match the pins of **r.r2** and **op42fz_2.u1**.

number	**gain=1,**	**# low frequency gain**
	fo=1k,	**# corner frequency**
	rin=1k	**# input impedance**

These are the declarations of the template arguments. There are three ways that arguments of the module template correspond to arguments of the underlying templates:

- Pass in directly—the value of an argument of the module template is used directly by one or more arguments of one or more underlying templates (**rin**).

- Use in expression—the value of an argument of the module template is used in an expression by one or more arguments of one or more underlying templates (**gain, fo**).

- Indirect correspondence—an argument of an underlying template uses an expression or argument value from another underlying template (as done for **r.r1**).

{

Begin template body (end of header declarations).

number   pi = 3.14159 # local declaration of a constant

Declare a constant called **pi** and assign it a value for use in subsequent calculations.

 # Netlist of filter module
r.r2 in sum   = rin

Resistor **r.r2** uses the argument value for **rin** passed in directly from **low_pass**.

r.r3 out sum = rnom=r.r2*gain

Resistor **r.r3** uses the argument value for **gain** (passed in from **low_pass**) in an expression with the argument from **r.r2**. Note that the gain factor was not a user-specifiable feature in the original netlist of this module on page 71.

r.r1 pos 0   = rnom(r.r2) * rnom(r.r3) / (rnom(r.r2)+rnom(r.r3))

Resistor **r.r1** uses expressions of arguments from **r.r2** and **r.r3** in the netlist instead of an argument value passed in from **low_pass**.

c.c1 out sum = 1/(2*pi*fo*rnom(r.r3))

Capacitor **c.c1** uses the argument value for **fo** (passed in from **low_pass**) in an expression with the argument from **r.r3**.

op42fz_2.u1 sum pos out 0 0 vee vcc

Op amp **op42fz_2** is a component model provided as a library template. This instance uses default argument values from the library. Connection points are specified as shown to implement **low_pass**.

}

End of template body.

Conclusion

The **low_pass** template (contained in the file **low_pass.sin**) contains a netlist of existing templates. However, the **low_pass.sin** file can be instantiated in a netlist as well. A user can then specify values for the arguments of **low_pass**, which are then used by its internal templates. For example, the following shows a typical netlist entry:

low_pass.f1 v6 fb vpos vneg = gain=1k, fo=1meg, rin=500k

CHAPTER 7

Basic Linear Devices

Following the basic process outlined in Chapter 5, this chapter begins to apply the principles of the MAST modeling language to writing templates for a few of the most basic electrical elements—a constant current source, a constant voltage source, a resistor, a capacitor, and an inductor. The model for each element is developed into a template by identifying usage requirements, dependent variables, independent variables, and parameters. This is done with a diagram, a description of the governing equations, and a discussion of each portion of the template as it implements the model.

7.1   Constant Current Source

A constant current source maintains a fixed current through its pins regardless of the voltage across its pins. Thus, its characteristic equation can be written as follows:

$$is = cur \tag{7-1}$$

where is is the current through the source and cur is the user-specified value this current is to maintain.

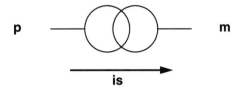

Figure 7-1. Constant current source.

A MAST template for this model is shown below (remember the # sign indicates a comment). Each line of the template is numbered as part of the comment and described below.

template isource p m = cur	# 1. Header line
electrical p, m	# 2. Connection point declaration
number cur=0	# 3. Argument declaration (set default to zero)
{	# 4. Begin template body
branch is = i(p->m)	# 5. Branch current declaration
is = cur	# 6. Equation for constant branch current
}	# 7. End template body

Line 1—The header line declares the name of the template (**isource**) and lists the connection points (**p, m**) along with the argument (**cur**). As described in Chapter 5, this line establishes the usage of the template.

Line 2—The connection points are declared as electrical pins **p** and **m**, which stand for plus and minus to indicate polarity. The word **electrical** means that **p** and **m** will use current as the through variable and voltage as the across variable. (Templates that contain netlists of other templates, as shown in Chapter 6, can "deduce" the connection point types from those declared in the underlying templates.)

Line 3—The argument **cur** is declared as a number and its default value is set to 0. A template user may specify any real number to override this default. The rule of thumb for a default argument value is that it should be chosen so that if the user does not assign a value, then the behavior modeled is not present. Here, a current source providing the

Basic Linear Devices

default amount of 0 amperes would effectively not be present in the circuit.

Line 4—The left brace, {, begins the body of the template (i.e., header, pin declarations, and argument declarations have been made).

Line 5—The branch current, is, is declared as the through variable (current) flowing through the branch between the pins **p** and **m** (as described in Chapter 5). The "arrow" operator, ->, indicates the direction of current flow.

Line 6—The branch current (is) is defined with the appropriate characteristic equation (**is=cur**). This is a relation that is computed; however, it is not an assignment that is satisfied through iterative solution by the simulator (i.e., it doesn't define a relationship that is satisfied through KCL).

Line 7—The right brace, }, ends the body of the template.

7.2 Resistor

According to Ohm's law, a resistor maintains a ratio between its current and voltage. As described in Chapter 5, the characteristic equation of a template should express the through variable as a function of the across variable for better simulation efficiency. Hence, Ohm's law is expressed with current (the through variable) as a function of voltage (the across variable), with the resistance as a model parameter:

$$ir = vr/res \tag{7-2}$$

where ir is the current through the resistor, vr is the voltage across the resistor, and res is the user-specified value of the resistor.

A MAST template for this simple resistor model is shown below. This template is quite similar to the constant current source template in Section 7.1. The only addition is the declaration of the branch voltage, **vr**. Note that the polarity of **vr** is determined by the pin order in the branch declaration (positive first). This branch declaration is equivalent to writing **vr = v(p) - v(m)**.

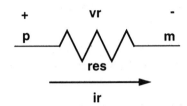

Figure 7-2. Resistor symbol.

```
template resistor p m = res      # Header line
electrical p, m                  # Connection point declaration
number res                       # Argument declaration (user must provide value)
{                                # Begin template body
     branch ir = i(p->m)         # Branch current declaration
     branch vr = v(p,m)          # Branch voltage declaration...vr = v(p) - v(m)
     ir = vr/res                 # Equation for branch current
}                                # End template body
```

Another difference between the resistor template and the current source template is that the resistor argument (**res**) is not assigned a default (initial) value in the template. In this case, the user is required to specify a value for **res** before the template can be used in simulation. This is principally because the writer cannot specify a reasonable default value for a resistor such that the template will have no effect in the circuit.

If constructs (condition checking)

Obviously, setting the value of **res** equal to zero in the above template could cause a problem in the template equation. In an actual template, the writer would want to check for special values (such as zero resistance) and take the appropriate action. To test whether the value of **res** has a valid value, an if construct can be used (see Appendix A). An if construct is a MAST statement that uses boolean operations to test for various conditions that can occur in a template. Such tests result in the value 1 if true and 0 if false.

Basic Linear Devices

The if expression is one such construct, having the following general syntax:

if (*conditional expression*) {*expressions1*}
else {*expressions2*}

This means that if the *conditional expression* is true, then *expressions1* are evaluated; if the *conditional expression* is false, then *expressions2* are evaluated. This could be used to detect whether zero has been specified for **res** and to do something about it if it has (i.e., bulletproofing). Note that condition checking is performed *before* calculation of the template equations.

The resistor template can be rewritten using an if expression to provide two different characteristic equations. The equation that gets used depends on whether the user specifies **res=0** or not. If the user specifies **res=0**, the first equation is in effect, using a local parameter (**xres**) instead of the argument (**res**). The value of **xres** has been arbitrarily set to $1\,m\Omega$ so that simulation can proceed. If the user specifies a nonzero value for **res**, the second equation is in effect.

```
template resistor p m = res
electrical p, m
number res
{
branch ir = i(p->m)
branch vr = v(p,m)
number xres=1m              # Local variable used in first equation (below)
    if(res==0) {
        ir=vr/xres          # First equation (if user sets res=0)
        }
    else {
        ir=vr/res           # Second equation (if user sets res~=0)
        }
}
```

7.3 Capacitor—Using Differentiation

A constant capacitor maintains a ratio between its charge (Q) and voltage (V). Its characteristic equation can be written as follows:

$$Q_c = C \cdot V_c \tag{7-3}$$

where C is the value of the capacitor. Noting that:

$$I = dq/dt \tag{7-4}$$

then Eq. (7-3) can be substituted into Eq. (7-4), which makes the following true and expresses i=f(v):

$$I_c = d(CV_c)/dt \tag{7-5}$$

The capacitor template is shown below. It introduces two new constructs:

- An intermediate variable, a **val**
- The differentiation operator, **d_by_dt**

template capacitor p m = cap	# Header line
electrical p, m	# Connection points
number cap=0	# Argument (set to default)
{	# Begin template body
branch ic=i(p->m)	# Branch current
branch vc=v(p,m)	# Branch voltage
val q qc	# Charge declaration (an intermediate variable)
qc = vc * cap	# Charge definition
ic = d_by_dt(qc)	# Characteristic equation
}	# End template body

Intermediate variables

A **val** can be thought of as a "scratch" variable. It does not require additional simulation time, but it can be used for intermediate calculations and can be displayed following simulation.

Basic Linear Devices

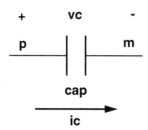

Figure 7-3. Capacitor symbol.

Although it is not necessary to declare a **val**, doing so provides the following advantages for variable usage:

- Easier to read
- Easier to maintain
- Easier to debug
- Easier to modify
- Provides appropriate axis labels when the curves are displayed (in this case, charge will be displayed with units of coulombs)

The general form for declaring a **val** is as follows:

> **val** *unit identifier [,identifier, ...]*

where *unit* is a predefined unit. The Saber simulator uses an include file that defines a wide variety of units for electrical, mechanical, magnetic, thermal, hydraulic, and optical variables.

Differentiation

In MAST, differentiation is accomplished with the **d_by_dt** operator. The only limitation to the use of the **d_by_dt** operator is that it can appear only in a summation, not products. For example, in order to express the following:

$$a = b \cdot dc/dt \tag{7-6}$$

it would be necessary to first bring **b** inside the derivative:

$$a = d(bc)/dt \tag{7-7}$$

which would expressed in a template equation as:

   a = d_by_dt(b*c).

Of course, this is only mathematically valid if **b** is a constant with respect to time. If **b** is not constant, then in order to implement Eq. (7-6), the result of **dc/dt** must be assigned to an intermediate **var** as follows:

   x: x=d_by_dt(c)
   a = b*x

7.4 Constant Voltage Source

As its name implies, a constant voltage source maintains a constant voltage across its pins regardless of the current through it. This is implemented by the following equation:

$$vs = vlt \qquad (7\text{-}8)$$

Because the voltage source cannot determine its own current, it does not assign a value to its branch current **is** as is done in the current source template. Consequently, the simulator must determine the current from the rest of the circuit that is connected to the voltage source. Nonetheless, because current does flow through this template from **p** to **m**, **is** must be declared as a branch current (i.e., it is a dependent variable for which the simulator must solve).

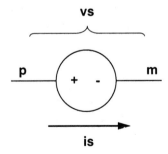

Figure 7-4. Constant voltage source.

Basic Linear Devices

Note that the branch declarations for the current and voltage can be shared. The comma (,) is used as a line continuation character in this case. This is very useful for commenting lists (as shown) and streamlining the code. Such enhancements help make the template easier to read, debug, modify, and maintain.

```
template vsource p m = vlt      # Header line
electrical p, m                 # Connection points
number vlt=0                    # Argument (set to default)
{                               # Begin template body
    branch is = i(p->m),        # Branch current
    vs = v(p,m)                 # Branch voltage
    vs = vlt                    # Characteristic equation
}                               # End template body
```

7.5 Inductor—Using Integration

The characteristic equation for an inductor can be written as:

$$i_L = \int V_L/L \, dt \tag{7-9}$$

Although this is in the preferred form of i=f(v), integration is not available explicitly in the MAST language. The approach is to differentiate both sides of the equation until there is no longer an integral:

$$di_L/dt = V_L/L \tag{7-10}$$

Notice that so far, all the characteristic equations in MAST have been of the form *variable = expression*. For the inductor, the characteristic equation can be formulated in either an assignment statement of this form, or the equation can be preceded by the reserved word **make** (see Chapter 5). Both implementations are shown below.

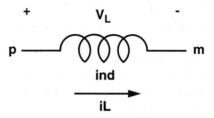

Figure 7-5. Inductor symbol.

The template that uses the assignment statement is listed first, followed by the template that uses **make**.

template inductor p m = ind	# Inductor not using "make"
electrical p, m	# Connection points
number ind=0	# Argument initialized to 0
{	# Begin template body
branch il=i(p->m),	# Branch current
vl=v(p,m)	# Branch voltage
vl = d_by_dt(il*ind)	# Characteristic equation
}	# End template body

template inductor p m = ind	# Inductor using "make"
electrical p, m	# Connection points
number ind=0	# Argument initialized to 0
{	# Begin template body
branch il=i(p->m),	# Branch current
vl=v(p,m)	# Branch voltage
make d_by_dt(il) = vl / ind	# Characteristic equation
}	# End template body

What *is* wrong with the second implementation?

CHAPTER 8

Piecewise Linear and Table Lookup Modeling

At times, instead of a characteristic equation, it may be desirable to base a model upon data or to approximate its behavior with a piecewise linear representation. This might apply to a behavioral model of a block of circuitry that performs a signal processing function or a semiconductor device model. Reasons that it may be desirable to create a table-based or piecewise linear model include:

- A device or circuit may not yet have a characteristic equation available. As the dimensions of transistors become smaller, it is typically the case that the development of physical models lags behind the processing development. An experimental data-driven model is one of the only means in which simulations can be performed at the transistor level with any measure of accuracy [1-3].

- Simulation speed—table-based models often simulate faster [4].

- Portions of the design or model may be specified in a piecewise linear (PWL) fashion, such as a nonlinear load.

- Modeling at abstract levels may be simpler with a data table approach, such as a piecewise linear diode that uses a simplified definition of turn-on voltage and on/off resistances. Another case is that data derived from a manufacturer's data sheet can be tabularized for a quick, simple model.

It is important that a modeling language consider these important aspects of modeling and make provisions for it. Foreign routines (in FORTRAN or C) are the mechanism by which table-driven models are realized in MAST. The MAST modeling language allows foreign routines to be called from within a template (see Appendix A). This provides substantial flexibility for choosing the data storage and interpolation algorithms most suited to the element being modeled. It also allows reusing code, since these algorithms can be shared among various software packages (such as between a simulator and a characterization tool).

MAST provides a piecewise linear interpolation algorithm (**pwl1**) for one-dimensional data tables. It serves as an illustrative tool for the development of other routines as well. Multi-dimensional piecewise linear interpolation has been implemented based on this foreign routine mechanism [5]. This allows table-based modeling to extend to MOSFETs and other semiconductor devices, as well as more complex models [6, 7].

To illustrate piecewise linear and table-based model implementation, a simple model of a voltage limiter (**vlim**) and a piecewise linear conductance (**pwlcond**) are described in the following sections.

8.1 Nonlinear Modeling

A linear model is characterized by the fact that its template equations include only linear functions of independent variables.[1] That is, there are no products or ratios of independent variables in a template equation, and no independent variable is an argument of a foreign or built-in function (except **d_by_dt** and **delay**).

---

[1] After substitution of all relevant expressions from **val** definitions.

If one or more of these requirements for a linear template is not met, the template is considered nonlinear. Note that a template can include nonlinear assignment statements yet still describe a linear element. The important question is whether the nonlinearities enter the template equations.

For example, a resistor template that defines power as the square of the voltage drop across a resistor divided by the resistance is still linear because power does not enter the template equation. In other words, power is part of the resistor template, but is not part of the resistor model (V=RI). *A template can also include nonlinear functions of time, frequency, or any parameter, without being a nonlinear template.*

Nonlinear models are not confined to curvilinear functions—other types include those whose outputs have discontinuities or regions of piecewise linear behavior. The examples in this chapter illustrate these kinds of nonlinear characteristics.

Appendix A describes some of the issues that can arise when modeling a nonlinear element. Most of these are handled automatically by the Saber simulator; however, there are MAST constructs that allow user-specification for more efficient simulation. All such constructs require statements in the control section of the template.

8.2 Modeling a Simple Voltage Limiter

The first example in this chapter shows how to use conditional expressions (if-else) in the template equation to define three regions of a symmetric voltage limiter shown in Figure 8-1. The template for this limiter, **vlim**, is shown below. This is an example of piecewise linear modeling without using data tables.

Characteristic equations

The characteristic or governing equations of the voltage limiter illustrated in Figure 8-1 are:

$V_{out} = -V_{max}$ if $V_{in} < -V_{max}$ (8-1)

$V_{out} = V_{in}$ if $-V_{max} \leq V_{in} \leq V_{max}$ (8-2)

$V_{out} = V_{max}$ if $V_{in} \geq V_{max}$ (8-3)

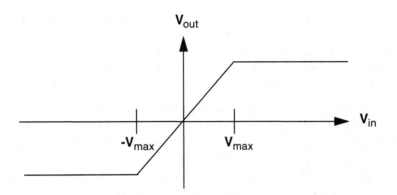

Figure 8-1. Ideal voltage limiter characteristics.

```
template vlim ip im op om = vmax
electrical ip,im,op,om
number vmax
{
branch  vin=v(ip,im)
branch  vout=v(op,om), iout=i(op->om)
struc {
     number bp, inc;
     } nvin[*]
number slope = 1u
vmx = abs(vmax)
nvin = [(-vmx,1.9*vmx),(vmx,0)]
control_section {
     newton_step(vin,nvin)
     }
# template equations using conditional expressions
vout =    if vin < -vmx then  -vmx + slope*(vin + vmx) \
          else    if vin > vmx then   vmx + slope*(vin - vmx) \
          else    vin
}
```

Template header and header declarations

The **vlim** template is a template with one argument, the limiting voltage (**vmax**). Its header and corresponding declarations are:

> template vlim ip im op om = vmax
> electrical ip, im, op, om
> number vmax

Note that there is no default value for **vmax**, which makes it mandatory for a user to specify an instance value. Note that this value may be specified as positive or negative—the template uses the absolute value of **vmax**, as described below.

Local declarations

The branch declarations (below) enable the characteristic equations to express the output voltage (**vout**) as a function of the input voltage (**vin**), while finding the current contribution (**iout**) required to make this true.

> branch vin=v(ip,im)
> branch vout=v(op,om), iout=i(op->om)

Although negative values for **vmax** are allowed, the equations for determining **vout** assume **vmax** is positive. That is, they use the absolute value of **vmax**, which is obtained by using **abs**, a built-in function for absolute value (**abs**). This absolute value of **vmax** is assigned to the local parameter **vmx** as follows:

> vmx = abs(vmax)

A variable named **slope** is declared as a local parameter and initialized to 10^{-6}. This is specified in the template as

> number slope=1u

Template equations

The template equations for **vlim** closely reflect the limiting characteristics given in Eqs. (8-1), (8-2), and (8-3), except that they include a nonzero slope in the limiting regions (**slope**), as shown in Figure 8-2. Specifying regions like this in an equation is efficiently accomplished by using an if expression.

An if expression is a compact form of one or more if statements. It can test a variety of conditions and then assign a value to a variable based on those conditions (see Appendix A). This is done for the template equation by making each of the three regions shown in Figure 8-1 a condition for the dependent variable, **vout**:

    vout = if vin < -vmx then -vmx + slope*(vin + vmx) \
        else if vin > vmx then vmx + slope*(vin - vmx) \
        else vin

Syntax

Note that this expression defines three conditions for the value of **vout**, although **vout** is explicitly listed only once. That is, in the first two lines, the value of **vout** evaluates to the expression following **then**. In the last line, the value of **vout** evaluates to the value of **vin**. Also note the use of the backslash (\) as a line continuation character.

Requirements for if constructs

There are several points worth noting about these conditional statements:

- If an independent variable, a branch variable, or a **val** that is a function of independent variables appear in the condition of an if statement, the variables defined in the body of the if statement depend nonlinearly on the variable used in the condition. In this example, **vout** depends nonlinearly on **vin**.

- Nonlinear models implemented with an if construct must be continuous from one region to the next. In these template equations, it is necessary to force continuity at **vin** = ±**vmax**. Discontinuities can cause problems during simulations (e.g., small time steps and long simulation times, or nonconvergence).

- Variables in an if construct must always defined, regardless of the conditions. One way to accomplish this is to make sure that any variable defined in any condition of an if construct is defined.

- A dependent variable defined in the body of an if construct should never be set to a constant value. The reason is that if the simulator, while iterating to find the solution of nonlinear equations, goes into a limiting region, it might not be able to get out of the region if the slope of the function is equal to 0—that is, the voltage limiter might latch.

To prevent this problem, a small but non-zero multiple of **vin** named **slope** is added to **vout** (as shown in Figure 8-2). In most cases, adding a very small slope yields a more realistic model than just a constant limit.

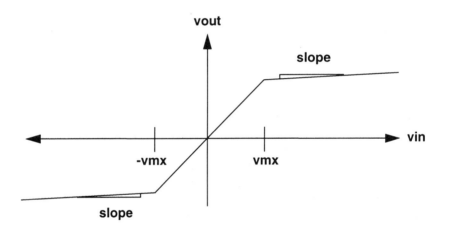

Figure 8-2. Voltage limiter template characteristics.

Control section

The **newton_step** statement is a MAST construct that interacts with the simulator for the purpose of making the model more efficient. Newton steps are described more thoroughly in Appendix A. For the present, it is sufficient to understand that the **newton_step** statement "assigns" the newton steps **nvin** to the independent variable **vin**.

8.3 Modeling a Nonlinear Conductance

The next example is that of a nonlinear conductance—a fairly common model used in simulation. The symbol of this device is given in Figure 8-2. The template is given below.

```
template pwlcond p m = vi_array      # Header
electrical p, m
struc {                              # Define a structure-- a new type of parameter
    number v, i                      # One number for voltage and one for current
    } vi_array[*]                    # Declare variable length array of structures
{
foreign pwl1                         # Declare foreign subroutine
branch ig=i(p->m), vg=v(p,m)
ig = pwl1(2, vi_array, vg)           # Given voltage and array, the pwl1 routine
                                     # determines current
}
```

This template defines conductance as the ratio of current to voltage at a series of points and along the line segments between those points. The user specifies each point as a pair of values—one for voltage and one for current—within a structure. The structure is then used as an array named vi_array, which is another complex data type. It is essentially a list of parameters that are of the same data type; in this case, the vi_array array is an array of structures (rather than being a single structure).

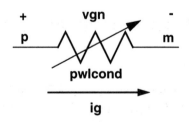

Figure 8-3. Piecewise linear conductance symbol.

An array does not need to be explicitly declared the way a number or a structure does. Here, the array of structures is declared by following the array name (vi_array) with the size of the array within square brackets ([]). The asterisk (*) declares the array to be of variable length. A variable-length array allows the user to specify any number of structures (i.e., pairs of voltage-current points) that define piecewise linear segments of conductance. Once voltage and current values have been specified, they are used by the foreign subroutine named **pwl1**, which is declared in the local declarations section of the template.

A subroutine may be declared in one of two ways:

- **foreign**
- **foreign number**

A subroutine declared as **foreign** cannot appear in expressions, but it can return more than one value. A subroutine declared as **foreign number** can only return one value, but can be used in an expression. In this template, either declaration would work for **pwl1**.

A foreign routine can serve several purposes. The first argument passed to the routine is a code that specifies the purpose of the call to the routine. In the case shown above (code=2), the **pwl1** routine is passed the characteristic array and the independent (x) coordinate. The **pwl1** routine then computes the associated dependent (y) coordinate. It uses linear interpolation if the independent coordinate is within the array values. Otherwise, it extrapolates from the first (or last) two points in the array.

The branch statement defines the conductor current **ig** to flow from pin **p** to pin **m** and the positive conductor voltage to be that of **p** with respect to **m**. Next, **pwl1** is called as described above.

As an example, suppose that the following data table of voltage-current pairs was to be used to model a piecewise linear conductance.

V	I
-10	-1
-5	-1
5	1
10	1

The following netlist instance could be used in a circuit for this device:

   pwlcond.1 node1 node2 = vi_array =[(-10,-1),(-5,-1),(5,1),(10,1)]

Conclusions

Using an AHDL allows one to create a model that is not implemented exclusively as a description of the ordinary differential equations that govern the behavior of a device. Although the same thought processes are involved, the flexibility of MAST permits specifying output behavior directly as piecewise linear regions, as well as using foreign routines to calculate data-based output. It is possible for such routines to provide external data from multi-dimensional tables back to the template.

8.4   References

1. A. Rofougaran and A. Abidi, "A Table Lookup FET Model for Accurate Circuit Simulation," *IEEE Trans. Computer-Aided Design*, Vol. 12, pp. 324-335, Feb. 1993.

2. T. Shima, T. Sugawara, S. Moriyama, and H. Yamada, "Three-Dimensional Table-lookup Model for Precise Circuit Simulation," *IEEE Journal of Solid-State Circuits*, Vol. SC-17, pp. 449-453, June 1982.

3. W. M. Coughran, E. H. Grosse, and D. J. Rose, "CAzM: A circuit analyzer with macromodeling", *IEEE Trans. Electron Devices*, vol. ED-30, pp. 1207-1213, Sept. 1983.

4. K. S. Yoon and P. E. Allen, "An adjustable accuracy model for VLSI analog circuits using lookup tables," *Journal of Analog Integrated Circuits and Signal Processing*, Kluwer Academic Publishers, Vol. 1, No. 1, pp. 45-64, Sept. 1991.

5. H. A. Mantooth, *Higher Level Modeling of Analog Integrated Circuits*, Ph.D. Dissertation, School of Electrical Engineering, Georgia Institute of Technology, Aug. 1990.

6. H. A. Mantooth, P. E. Allen, "Behavioral Simulation of a 3-bit Flash ADC," *IEEE Proc. of Int. Symposium on Circuits Syst.*, Vol. 2, pp. 1356-1359, May 1990.

7. H. A. Mantooth, P. E. Allen, "A Higher Level Modeling Procedure for Analog Integrated Circuits," *Journal of Analog Integrated Circuits and Signal Processing*, Vol. 3, pp. 181-195, May 1993.

CHAPTER 9

Modeling Nonlinear Devices

In this chapter, the discussion of modeling various classes of functions turns to that of nonlinear devices. Although the examples presented are for semiconductor device models, the concepts can be extended to modeling arbitrary nonlinear functions. The examples are simplified models of the diode and MOSFET. The thought process described in Chapters 2 and 5 is once again used to develop the models as if creating them for the first time. This thought process consists of the following major steps:

- Determine how the device interfaces to the elements to which it could be connected
- Relate the governing equations to a large-signal model
- Determine the required algebraic relationships
- Parameterize the equations

In each of the examples presented, the first step is obvious from the nature of the device. Both the diode and MOSFET are electrical devices requiring analog connections points.

9.1 Modeling a p-n Junction Diode

When modeling a nonlinear device for the first time, it may be the case that the governing equations have been developed and that the modeling task is more one of implementation. On the other hand, it may be that the equations must first be derived. In the first case, the starting point is obvious. The governing equations must be related to one another for the purpose of implementing a model. Thus, the next major step is to represent these equations pictorially in the form of a large-signal topology. This topology captures the interrelationship of equations that is indispensable for models used by an analog simulator. However, if the equations are unknown, then it helps to create a large-signal topology that, once the appropriate equations are derived, will model the desired phenomena.

Diode equations

For the model of a p-n junction diode, the governing equations are already available [1]. The governing equation for current flow in a p-n junction diode is as follows:

$$I_d = I_s(e^{V_d/n \cdot V_t} - 1) \tag{9-1}$$

where $V_t = kT/q$, n is the emission coefficient, I_s is the saturation current, and V_d is the junction diode voltage.

The governing equation for the charge in the diode is:

$$Q_{diode} = Q_{diff} + Q_{depl} \tag{9-2a}$$

where Q_{diff} is the charge stored due to diffusion and Q_{depl} is the charge in the depletion region. The relationships [2] for each of these components of charge in the diode are given by:

$$Q_{diff} = \tau \cdot I_d \tag{9-2b}$$

Modeling Nonlinear Devices

For $V_d < fc \cdot V_j$,

$$Q_{depl} = V_j \cdot C_{jo} \cdot \frac{\left[\left(1 - \frac{V_d}{V_j}\right)\right]^{1-m}}{1-m}$$

For $V_d \geq fc \cdot V_j$,

$$Q_{depl} = C_{jo} \cdot V_j \cdot \frac{1 - (1-fc)^{(1-m)}}{1-m} + \left[\frac{C_{jo}}{(1-fc)^{(1+m)}}\right] \cdot$$

$$\left[1 - (fc \cdot (1+m)) \cdot (V_d - (fc \cdot V_j)) + \frac{m \cdot (V_d^2 - (fc \cdot V_j)^2)}{(2 \cdot V_j)}\right]$$

(9-2c)

The large-signal topology

Figure 9-1 shows the large-signal model of the diode. The current source symbol represents the contribution from Eq. (9-1). The nonlinear charge symbols in parallel with the current source represent the contribution of Eqs. (9-2). A series resistance (R_s) is added as well. Note that the voltage that controls the diode current, I_d, and the nonlinear charge storage is V_d, the voltage across this dependent source. The total diode voltage, V_{pn}, includes the voltage drop across the linear series resistance.

While it is obvious in this simple example how the governing equations are related to one another, the large-signal topology provides the relational details necessary to implement the model. This is true whether the model is written as part of an analog simulator or with an analog HDL. The detail of whether to "connect" the nonlinear charge to the internal node or the external node of the diode is one such detail that makes a significant difference in the functionality of the model.

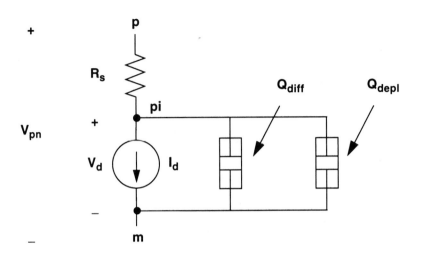

Figure 9-1. Large-signal topology of p-n junction diode.

Diode parameters

Since this model is fairly simple, there is no algebraic manipulation required to implement the equations. Thus, the final step in the modeling thought process is the parameterization of the governing equations. For optimum usability, this is done so that the model can be characterized based on electrical measurements of the device or from data sheet information (if sufficient information is published).

The traditional parameters that arise from Eq. (9-1) are:

n	emission coefficient
I_s	reverse saturation current
T	junction temperature

The parameters that arise from Eqs. (9-2) are:

V_j	built-in junction potential
fc	empirical constant for straight line approximation on capacitance
C_{jo}	zero-bias junction capacitance
m	junction grading coefficient
τ	transit time (equal to recombination lifetime of minority carriers here)

Diode template

When implementing the model in MAST, the thought process now reverses. First, the parameters are created (along with header, header declarations, comments, etc.), then the equations are formulated, and then the large-signal topology is implemented. Constants, local variables, and parameter checking is also included. The MAST template[1] of this example diode is shown below.

```
template diode p n = model, ic        # Header
electrical p, n                       # Header Declarations
struc {                               # Model structure
    number  n = 1,
            is = 10f,
            tau = 0,
            cjo = 0,
            m = .5,
            fc = .5,
            vj = 1,
            rs = 0
    } model=()
external number temp                  # Global temperature variable
number ic=undef                       # Initial condition on junc. voltage
{
<consts.sin
electrical pi                         # Internal node
# Local Declarations
number  wn,                           # Working value of n
        vt,                           # Thermal voltage
        fcvj                          # model->fc*model->vj
val q   qdiode                        # Charge in diode
val i   id                            # DC current through diode
val v   vd,                           # Junction voltage
        vpn                           # Total diode voltage
```

---

[1] Refer to Chapter 5 and Appendix A for general information on template structure and organization.

```
parameters{                                 # Parameters Section
    vt = math_boltz*(temp+math_ctok)/math_charge
    fcvj = model->fc*model->vj
    if(model->n <= 0)         wn = .1
    else                      wn = model->n
    }
values{                                     # Values Section
    vpn = v(p) - v(n)
    vd = v(pi) - v(n)
    id = model->is*(limexp(vd/(wn*vt)) - 1)
    if (vd < fcvj)            {
        qdiode = model->tau*id + model->vj*model->cjo*
                (1 - (1 - vd/model->vj)**(1-model->m))/(1 - model->m)
                            }
    else {
        qdiode = model->tau*id + model->cjo*model->vj*(1 -
                (1 - model->fc)**(1 - model->m))/(1 - model->m) +
                (model->cjo/(1 - model->fc)**(1 + model->m)) *
                (((1 - model->fc*(1+model->m)))* (vd - fcvj) +
                model->m/(2*model->vj)*(vd**2 - fcvj**2) )
        }
    }
control_section{                            # Control Section
    if(model->rs == 0) collapse(p,pi)
    #...Assign newton steps here (see Appendix A)
    newton_step(vd, [0,0.1,2,0])
    #...Small signal parameters of diode
    device_type("diode", "example")
    small_signal(rc,resistance,"series resistance",model->rs)
    ss_partial(g_d, id, vd)
    small_signal(req,resistance,"pi-n resistance",
                    if g_d ~= 0 then 1/g_d else inf)
    small_signal(cj,capacitance,"pi-n capacitance", qdiode, vd)
    #...initial condition on junction voltage
    initial_condition(vd,ic)
    }
equations{                                  # Equations Section
    i(pi->n) += id + d_by_dt(qdiode)
    if(model->rs ~= 0)    i(p->pi) += v(p,pi)/model->rs
    }
}
```

Modeling Nonlinear Devices **103**

As mentioned in Chapter 5, explicit template sections (that is, those within the opening and closing braces) may appear in any order. The code shown above is one possible implementation of these diode equations. It is a straightforward realization of the governing equations and flows from declarations to algebraic relations to the final differential equations.

Header and header declarations

The diode template begins with the header information. The template is named **diode**. It has external connection points, **p** and **n**. It has two arguments—the first (**model**) is a set of basic diode model parameters; the second (**ic**) is the initial condition on the junction voltage. The **model** argument is a structure that contains the diode parameters described on page 100. The header describes how the diode model would be used in a netlist. One such instance is

> diode..model dmod=(is=15f, n=1.1, rs=1m)
> diode.1 pos neg = model=dmod, ic=0.65

The first statement is the definition of the model parameters for this instance. It is similar to a .MODEL statement in a SPICE netlist. The second statement is an instance of the **diode** template, which is connected from node **pos** to node **neg**.

Temperature (**temp**) is an externally defined simulation constant. Although it is not a template argument, **temp** (a reserved word) is defined globally for all simulation models. It can be changed locally for a given instance in a design.

Local declarations

The next section of the template is the local declarations. Some predefined constants are made available by using a separate file named **consts.sin** as an include file (see Appendix A). Among the constants provided in this file are Boltzmann's constant (**math_boltz**) and the value of electronic charge (**math_charge**). An internal electrical node (**pi**) is defined for the connection between the series resistance and the current source in the large-signal topology of Figure 9-1. The other declarations are intermediate numbers and values used in the parameters and values section (described below).

The local declarations are typically included in the code as they are created in their respective sections.

Parameters section

The parameters section is where one-time calculations are performed to create the locally declared numbers as a function of the input arguments (the **model** structure in this case). An example of this in the **diode** template is the computation of the thermal voltage, **vt**. Although **vt** depends on temperature, temperature does not change during a simulation. Also, this section is where all argument checking is performed to ensure that valid values are used. Only one such check has been included in this implementation. The value of **n** is tested for being greater than zero. In a complete implementation, all remaining arguments would be similarly checked in this section.

Values section

The values section contains the governing algebraic equations of Eqs. (9-1) and (9-2). The voltages **vpn** and **vd** are defined in terms of the pin voltages. Note that **vpn** is provided purely for observation, since it is not actually used in the calculation of the diode current or charge. There is no simulation penalty for including this term, since it does not contribute to the equations ultimately formulated in the equations section below. Another such term that could be included for observation is the power dissipation as calculated from the junction current and voltage and the series resistance:

   pd = (vpn-vd)**2/model->rs + id*vd

Equations section

Although the control section appears next in the **diode** template, the next implementation step is the equations section. This is where the algebraic governing equations are "connected" together as indicated in the large-signal topology. In this case, it is straightforward to sum the current **id** and the derivative of the charge **qdiode** from nodes **pi** to **n**. In the event that the series resistance is nonzero, the current (expressed as a function of the voltage difference) through the resistance is also included.

Control section

The control section of the **diode** template contains a new construct—the **collapse** statement. This is used to collapse two nodes into one if the resistance between them is zero. This improves computational efficiency of the model when possible, since each node requires an independent variable in the system matrix.

The next statement is the **newton_step** statement, which is a convergence aid covered in Appendix A. The next group of statements are those required for the output of the small-signal parameters of the diode model. These statements are executed as a post-processing step, which can be performed following a DC operating point analysis. These parameters include series resistance, effective resistance of the current source, and capacitance of the diode at the previously calculated operating point. Figure 9-2 shows an example of the output generated using these statements. The last statement in the control section sets the initial condition on the junction voltage (**vd**) of the diode. This statement has no effect if the value of the **ic** argument is equal to **undef**.

Part:	/diode.1		
Device type:	**diode**		
Parameter	*Name*	*Classification*	*Value*
series resistance	rc	resistance	0.001
pi-n resistance	req	resistance	4.53
pi-n capacitance	cj	capacitance	222n

*Figure 9-2. Small-signal parameters report from the **diode** template.*

Simplified template form

In preceding chapters, the simplified form of MAST has been employed. However, this template was written using explicit sections (i.e., values, equations, parameters, etc.) in order to emphasize some of the concepts required for robust model implementation of a nonlinear device. In a model such as the MOSFET in Section 9.2, it becomes more evident that explicit sections are helpful in accounting for the equations.

9.2 Modeling a MOSFET

As with the diode model, the process begins with known governing equations. In this example a version of the Shichman-Hodges MOS model [3] will be implemented. This is also commonly referred to as the Level 1 MOSFET model, which refers to the SPICE level that invokes this model.

MOSFET equations

MOSFET equations for the three regions of drain-source current (I_{ds}) are given below.

$Cutoff\,(V_{gs} - V_T \leq 0)$

$$I_{ds} = 0 \qquad (9\text{-}3a)$$

$Triode\,(V_{gs} - V_T > V_{ds}\text{ and }V_{gs} - V_T > 0)$

$$I_{ds} = \beta\left[V_{ds}(V_{gs} - V_T) - \frac{1}{2}\left(V_{ds}^2\right)\right] \qquad (9\text{-}3b)$$

$Saturation\,(V_{gs} - V_T \leq V_{ds}\text{ and }V_{gs} - V_T > 0)$

$$I_{ds} = (\beta/2)(V_{gs} - V_T)^2 \qquad (9\text{-}3c)$$

where V_T is the threshold voltage and $\beta = (kp\,(w_{eff}/l_{eff}))\cdot(1 + \lambda \cdot V_{ds})$

The governing equations for the bulk-drain and bulk-source diodes are similar to those given for the diode model in Section 9.1. However, these diodes contain no series resistances.

The governing equations for the intrinsic MOSFET charge (commonly referred to as the Meyer model) are *not* as implemented from SPICE2G.6 but have been taken from other work [4]. These charge equations are given for the four regions of operation listed below.

Modeling Nonlinear Devices

Below Flatband ($V_{gb} \leq V_{fb}$)

$$Q_{gate} = C_{ox}(V_{gb} - V_{fb}) \tag{9-4a}$$

$$Q_{bulk} = -Q_{gate} \tag{9-4b}$$

$$Q_{drain} = 0 \tag{9-4c}$$

$$Q_{source} = 0 \tag{9-4d}$$

Below Threshold ($V_{gdt} \leq V_{gst} \leq 0$)

$$Q_{gate} = C_{ox}\gamma \left[\sqrt{\frac{\gamma^2}{2} + V_{gb} - V_{fb}} - \frac{\gamma}{2} \right] \tag{9-5a}$$

$$Q_{bulk} = -Q_{gate} \tag{9-5b}$$

$$Q_{drain} = 0 \tag{9-5c}$$

$$Q_{source} = 0 \tag{9-5d}$$

Saturation region ($V_{gdt} \leq 0$ and $V_{gst} > 0$)

$$Q_{gate} = C_{ox}\{2/3\, V_{gst} + [V_{gst} + (V_T - V_{bi})]\} \tag{9-6a}$$

$$Q_{bulk} = -C_{ox}(V_T - V_{bi}) \tag{9-6b}$$

$$Q_{drain} = (-4/15)\, C_{ox}(V_{gst}) \tag{9-6c}$$

$$Q_{source} = (-2/5)\, C_{ox}(V_{gst}) \tag{9-6d}$$

Triode region ($V_{gdt} > 0$ and $V_{gst} > V_{gdt}$)

$$Q_{gate} = C_{ox}\{V_T - V_{bi} + 2/3\left[V_{gsdt} + V_{gst} - \frac{V_{gst}V_{gdt}}{V_{gst} + V_{gdt}}\right]\} \quad (9\text{-}7a)$$

$$Q_{bulk} = -C_{ox}(V_T - V_{bi}) \quad (9\text{-}7b)$$

$$Q_{drain} = -\frac{C_{ox}}{3}\left[\frac{1}{5}V_{gdt} + \frac{4}{5}V_{gst} + V_{gdt}\left(\frac{V_{gdt}}{V_{gsdt}}\right) + \frac{1}{5}\frac{V_{gst}V_{gdt}(V_{gdt} - V_{gst})}{(V_{gsdt})^2}\right]$$

$$(9\text{-}7c)$$

$$Q_{source} = -\frac{C_{ox}}{3}\left[\frac{1}{5}V_{gst} + \frac{4}{5}V_{gdt} + V_{gst}\left(\frac{V_{gst}}{V_{gsdt}}\right) + \frac{1}{5}\frac{V_{gst}V_{gdt}(V_{gst} - V_{gdt})}{(V_{gsdt})^2}\right]$$

$$(9\text{-}7d)$$

where $V_{bi} = V_{fb} + 2\phi_f$, $V_{gsdt} = V_{gst} - V_{gdt}$, $V_{gst} = V_{gs} - V_T$, and $V_{gdt} = V_{gd} - V_T$

The large-signal topology

Figure 9-3 shows the large-signal topology of the MOSFET. The model consists of three current generators: one for the drain-source current (I_{ds}) and two for the bulk diodes (I_{bd} and I_{bs}). There are six nonlinear charge functions in this model: one for each terminal (Q_{gate}, Q_{drain}, Q_{source}, and Q_{bulk}) and one for each bulk diode (Q_{bsd} and Q_{bdd}). There are three linear capacitance terms associated with the overlap capacitances (C_{gdo}, C_{gso}, and C_{gbo}). Finally, there are the series resistances on the drain (R_d) and the source (R_s).

Modeling Nonlinear Devices

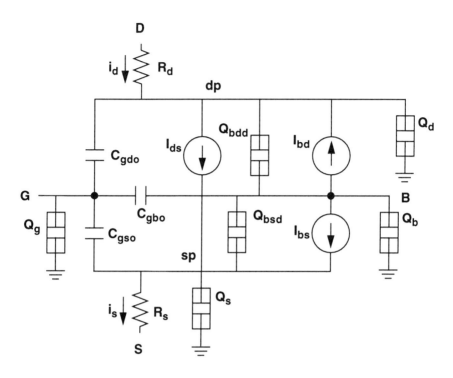

Figure 9-3. Large-signal topology of MOSFET model.

MOSFET parameters

The parameters chosen from the preceding equations that allow user-specifiable values are listed below. There are no new parameters for the improved Meyer model as compared to the original version.

DC current equations parameters

w	channel width
l	channel length
kp	transconductance
cox	oxide capacitance
lambda	channel-length modulation

Threshold voltage parameters (V_T)

vto	threshold voltage
gamma	bulk threshold (body effect)
phi	surface potential

Series resistance parameters

rs	source resistance
rd	drain resistance

Bulk diode parameters

is	saturation current
pb	junction potential
cj	bottom junction capacitance
mj	grading coefficient
cjsw	sidewall junction capacitance
mjsw	sidewall grading coefficient
fc	forward-bias junction capacitance coefficient

MOSFET template

The same steps that were followed for the diode are again performed for the MOSFET. However, due to the complexity of this model, the MOSFET template takes advantage of external MAST functions, as described in Appendix A. These functions are contained in files that are separate from the file of the MOSFET template. In addition, a "companion" template is created to declare all MOSFET parameters, so that they can be used by the main template and by the three MAST functions. Thus, there are five files used to implement the MOSFET model:

- **mos.sin** (the main MOSFET template)
- **mostype.sin** (the companion template)
- **mosparam.sin** (a MAST function)
- **mosval.sin** (a MAST function)
- **mosdiode.sin** (a MAST function)

These files will be described as the mos template is described. However, Figure 9-4 shows a block diagram of the relationship of these files to the main MOSFET template mos, which is given

Modeling Nonlinear Devices **111**

following the figure. Following the **mos** template is the source listing for the companion template, **mostype**. The source listings for the MAST functions are provided in Appendix B.

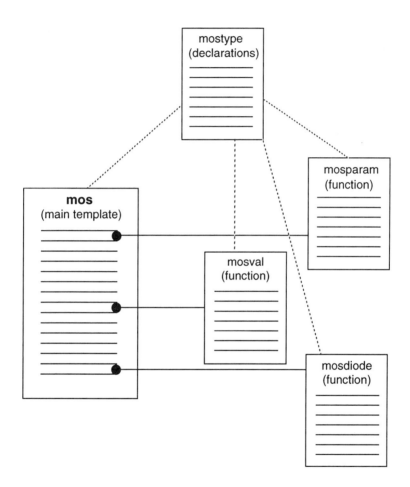

Figure 9-4. Block diagram of overall MOSFET model.

```
template mos d g s b = model, l, w, ic
electrical d,g,s,b                              # Electrical node declarations
number  l = 1u,
        w = 1u
mostype..model model=()                         # Declaration of model from mostype
external number temp
number ic[3] = [undef,undef,undef]              #...vds,vgs,vbs
{
electrical dp,sp                                #...internal nodes
mostype..work work=()                           # Declaration of work from mostype
val v    vg, vd, vs, vb, vsp, vdp, vgs, vsg, vgb, vds, vbs,
         vgdi, vbdi, vgsi, vdsi, vbsi, pxvgsi, pxvgdi, pxvdsi,
         pxvbsi, pxvbsid, pxvbdid
val i    ids, ibd, ibs, id, is
val q    qg, qb, qd, qs, qbsd, qbdd, qgso, qgdo, qgbo
parameters{
    work = mosparam(model,temp,l,w)
    }
values{
    #...Definition of pin and branch voltages
    vg = v(g)
    vd = v(d)
    vs = v(s)
    vb = v(b)
    vsp = v(sp)
    vdp = v(dp)
    vgs = vg - vs
    vsg = vs - vg
    vgb = vg - vb
    vds = vd - vs
    vbs = vb - vs
    vgdi = vg - vdp
    vbdi = vb - vdp
    vgsi = vg - vsp
    vdsi = vdp - vsp
    vbsi = vb - vsp
```

Modeling Nonlinear Devices

```
        #...Voltages required for the subroutine
        if (work->wtype > 0) {
            pxvgsi = vgsi
            pxvgdi = vgdi
            pxvdsi = vdsi
            pxvbsi = vbsi
            pxvbsid = vbsi
            pxvbdid = vbdi
            }
        else {
            pxvgsi = -vgsi
            pxvgdi = -vgdi
            pxvdsi = -vdsi
            pxvbsi = -vbsi
            pxvbsid = -vbsi
            pxvbdid = -vbdi
            }
        #...Call mosval to get current and charges
        (ids,qg,qb,qd,qs) = mosval(work,model,pxvgsi,pxvdsi,pxvbsi)
        #...Call mosdiode to get currents and charges for bulk diodes
        (ibs, qbsd) = mosdiode(work,model,pxvbsid)
        (ibd, qbdd) = mosdiode(work,model,pxvbdid)
        #...Calculate charges on overlap caps
        qgbo = work->wcgbe*vgb
        qgdo = work->wcgde*vgdi
        qgso = work->wcgse*vgsi
        }
control_section{
        #...Remove unneeded internal nodes:
        if (model->rd == 0) collapse (d,dp)
        if (model->rs == 0) collapse (s,sp)
        #...Define initial conditions
        initial_condition(vds,ic[1])
        initial_condition(vgs,ic[2])
        initial_condition(vbs,ic[3])
        #...Assign small signal parameters
        device_type("mosfet","example")
        ss_partial(ggs_d,ibs,vbsi)
        ss_partial(ggd_d,ibd,vbdi)
        ss_partial(gds_d,ids,vdsi)
```

```
        small_signal(gm,transconductance,"d(ids)/d(vgs)",ids,vgsi)
        small_signal(gmb,transconductance,"d(ids)/d(vbs)",ids,vbsi)
        small_signal(rd,resistance,"drain resistance",model->rd)
        small_signal(rs,resistance,"source resistance",model->rs)
        small_signal(rds,resistance,"d-s resistance",
                        if gds_d ~= 0 then 1/gds_d else undef)
        small_signal(rbs,resistance,"b-s resistance",
                        if ggs_d ~= 0 then 1/ggs_d else undef)
        small_signal(rbd,resistance,"b-d resistance",
                        if ggd_d ~= 0 then 1/ggd_d else undef)
        small_signal(cgs,capacitance,"-d(qg)/d(vs)",-qg,vsp)
        ss_partial(cgdi,-qg,vdp)
        ss_partial(cdgi,-qd,vg)
        small_signal(cgd,capacitance,"-d(qg)/d(vd)",cgdi)
        ss_partial(cgbi,-qg,vb)
        ss_partial(cbgi,-qb,vg)
        small_signal(cgb,capacitance,"-d(qg)/d(vb)",cgbi)
        small_signal(cbs,capacitance,"-d(qb)/d(vs)",-qb,vsp)
        ss_partial(cbdi,-qb,vdp)
        ss_partial(cdbi,-qd,vb)
        small_signal(cbd,capacitance,"-d(qb)/d(vd)",cbdi)
        small_signal(cm,capacitance,"cdg - cgd",cdgi - cgdi)
        small_signal(cmb,capacitance,"cdb - cbd",cdbi - cbdi)
        small_signal(cmx,capacitance,"cbg - cgb",cbgi - cgbi)
        small_signal(cbde,capacitance,"b-d capacitance",qbdd,vbdi)
        small_signal(cbse,capacitance,"b-s capacitance",qbsd,vbsi)
        small_signal(cgso,capacitance,"g-s overlap",qgso,vgsi)
        small_signal(cgdo,capacitance,"g-d overlap",qgdo,vgdi)
        small_signal(cgbo,capacitance,"g-b overlap",qgbo,vgb)
    }
equations{
    i(dp->sp) += ids
    i(b->dp) += ibd + d_by_dt(qbdd)
    i(b->sp) += ibs + d_by_dt(qbsd)
    i(g->sp) += d_by_dt(qgso)
    i(g->b) += d_by_dt(qgbo)
    i(g->dp) += d_by_dt(qgdo)
    i(g) += d_by_dt(qg)
    i(b) += d_by_dt(qb)
```

```
        i(d) += d_by_dt(qd)
        i(s) += d_by_dt(qs)
        if(model->rd ~= 0)   i(d->dp) += v(d,dp)/model->rd
        if(model->rs ~= 0)   i(s->sp) += v(s,sp)/model->rs
        }
}
```

Template listing for **mos** *template.*

```
template mostype = model, work
struc {
    enum {_n,_p} type=_n
    number  vto=1, kp=20u, gamma=0, phi=0.6, lambda=0, tox=100n,
            rd=0, rs=0, is=1.0e-14, pb=0.8, cgso, cgdo, cgbo,
            cj=0.0, mj=0.5, fc=.5, gmin=1p, ld=0, wd=0, tnom=27
    } model
struc {
    number  wtype, wvto, wkp, wgamma, wphi, wlambda,
            wcox, wbeta, wis, wvj, wcj, wfc, wvt, wcgbe,
            wcgde, wcgse, wleff, wweff, wfunc2
    } work
{}
```

Template listing for **mostype** *companion template.*

Header and header declarations

The **mos** template has four electrical connection points (**d, g, s, b**) and four arguments (**model, l, w, ic**). The **model** structure argument is declared in **mostype** (see above). The remaining three arguments are declared in the **mos** template. Temperature is once again declared as an external number, **temp**. From the header information the netlist usage is obvious. An example of this is:

   mostype..model nmos=(vto=1.1, kp=14u, cjo=1n)
   mos.1 drain gate source bulk = model=nmos, w=10u, l=10u

The first line is the model information for this instance. The second is the actual netlist entry for using the **mos** template.

The initial conditions for the **mos** model are defined in a three-element array. The initial conditions could be chosen to be any of the independent variables. In this case the initial conditions are imposed on three branch voltages **vds**, **vgs**, **vbs**, respectively. This assignment is made in the control section.

Local declarations

There are two internal nodes in the MOSFET model (**dp**, **sp**). These accommodate the series resistances of the drain and source shown in Figure 9-3, respectively. A **work** structure is referenced in the local declarations section. This is declared in the companion template **mostype**. This structure is the output of the MAST function **mosparam**. The "working" values of the model parameters are retained in this structure after they've been validated. The other declarations are the voltages, currents, and charges required in the values section. There are many more of these than are absolutely required to implement the model, but they are provided as additional values for monitoring.

Parameters section

The parameters section of **mos** is simply a call to the MAST function **mosparam**. This function processes all of the model arguments by checking for allowable ranges of values based on numerical and physical considerations. Model entities that are a function of model parameters and temperature are calculated in this function. An example of this is the temperature variation of the threshold voltage. Given a temperature, this function calculates the threshold voltage value at that temperature based on the value provided at 27°C and the temperature coefficients. This function as listed in Appendix B contains many representative computations, but is not intended to be exhaustive.

The use of the MAST function **mosparam** allows the code to be conveniently partitioned and object-oriented, in a sense. As a result the template appears simpler, containing more of the necessary information that a template requires.

Modeling Nonlinear Devices

Values section

As in the case of the diode model in the previous section of this chapter, the values section of the **mos** template contains the algebraic relations for the elements of the large-signal topology of Figure 9-3. It is composed of the definition of the various branch voltages including the adjusted values based on the type of MOSFET (nMOS or pMOS) being simulated. This is achieved with the *if* statement based on the type parameter which is **work->wtype** after being returned from the **mosparam** function. Next, the currents and charges are computed in the MAST functions **mosval** and **mosdiode**. The last expressions in the values section are those that calculate the charges on the overlap capacitances.

The primary governing equations of the MOS model are contained in the **mosval** function. This includes the drain-source current expressions of Eq. (9-3) and the charge expressions of Eqs. (9-4)-(9-7).

The **mosdiode** function contains the bulk diode model, which is quite similar to that already presented in Section 9.1.

Equations section

Once again, the equations section of the template is described next, as it follows from the work performed to create the values section. The equations section is a straightforward implementation of the large-signal topology of Figure 9-3. The drain-source current flows between the internal nodes **dp** and **sp**, respectively. The bulk diode terms (current and charge) are implemented between the drain-bulk and source-bulk as they were with the diode model. The capacitances are implemented as the derivatives of the charge terms calculated in the values section. This is also true of the intrinsic charges of the MOSFET (**qg, qd, qb, qs**). The series resistances are handled in a similar fashion as in the diode model.

Control section

The control section consists of much of the same entities as did the **diode** template:

- the **collapse** statement depending on the value of the series resistances

- the **initial_condition** statements assign initial values to V_{ds}, V_{gs}, and V_{bs}
- the **small_signal** statements make conductances and capacitances available for the small-signal parameters report [5].

Operational characteristics

Figure 9-5 shows a family of curves obtained by sweeping the drain-source voltage of the MOSFET at various values of the gate-source voltage.

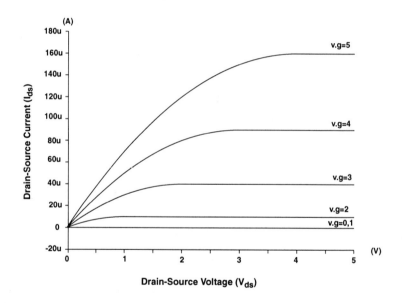

Figure 9-5. Output family of curves for MOSFET model.

Conclusions

In the same manner in which these two semiconductor device models were developed, any nonlinear behavior that is based on differential equations can be modeled using MAST. Other examples of nonlinear modeling are provided in subsequent chapters.

Modeling Nonlinear Devices **119**

As with any computational program, the efficiency of the resulting model is dependent upon the implementation. When it comes to modeling nonlinear behavior, simply minimizing the number of multiplications and divisions isn't sufficient. It is often necessary to use knowledge about the nature of the model to aid the simulator in finding solutions. One such construct provided by MAST is the newton step, which is described in detail in Appendix A.

9.3 References

1. B. G. Streetman, *Solid State Electronic Devices*, Prentice-Hall, pp. 172-173, 1980.

2. P. Antognetti and G. Massobrio, *Semiconductor Device Modeling with SPICE*, Chapter 1, McGraw-Hill, 1988.

3. H. Shichman and D. A. Hodges, "Modeling and simulation of insulated-gate field-effect transistor switching circuits," *IEEE Journal of Solid-State Circuits*, SC-3, 1968.

4. K. A. Sakallah, Y. T. Yen, S. S. Greenberg, "First-Order Charge Conserving MOS Capacitance Model," *IEEE Trans. on Computer-Aided Design*, Vol. 9, No. 1, pp. 99-108, January 1990.

5. Y. P. Tsividis, *Operation and modeling of the MOS transistor*, chapter 9, McGraw-Hill, New York, NY, 1987.

CHAPTER 10

Digital Modeling

This chapter introduces a new area of application for an AHDL—providing models that use a restricted set of discrete values that change only at discrete, scheduled time points. This type of modeling (referred to as *event-driven*) represents a significant departure from previous applications, both in the capability of a modeling language and in using this capability to write models.

Digital modeling is a specific type of event-driven modeling; as such, it requires a different set of MAST constructs and variables. The simple examples in this chapter are meant to introduce these, along with a feel for the approach in writing digital models.

10.1 Digital Characteristics

All the models presented in preceding chapters have been completely analog. Analog simulation has the following characteristics:

- All system variables are solved for simultaneously at each iteration.
- All variables can assume any value from a continuous range of values.

- Signals change values continuously over time.

Digital simulation was developed to reflect the differences between digital circuitry and analog circuitry. As such, digital simulation has the following characteristics:

- Signals change instantaneously at specific (discrete) times.
- System variables can only take certain values (known as states or logic levels). Among the most commonly used values for states are the following:
  - HIGH (logic 1, or true)
  - LOW (logic 0, or false)
  - unknown (X, derived from conflicts between 1 and 0 or from uninitialized variables)

In addition to these three values that system variables can hold (0, 1, X), models may also use other states, such as "high impedance" (Z, or inactive). This state is necessary for implementing buses, where more than one model may place a signal on a node, but only one signal may be present at a time. MAST uses a set of four state values, called logic_4 values, that correspond to these digital states (l4_1, l4_0, l4_x, l4_z).

Digital signals change to these state values instantaneously (i.e., at discrete time points instead of over a continuous interval of time). Therefore, a digital simulator only needs to keep track of the timing of these changes (referred to as *events*) and the value held by a node between events. To accomplish this most efficiently, digital simulators use what is known as an *event queue*.

An event queue is essentially a time-ordered list of events to be processed. "Digital time" progresses by taking each event off the queue in order, placing the changed state on a variable, and activating the models that are "listening" for such an event. In MAST, models "listen" for events using the **when(event_on())** statement, and place events on the queue with a **schedule_event()** statement.

10.2 Digital Inverter Template

Figure 10-1 shows an example of a very simple inverter model, whose template is shown below.

template inverter in out	# Header line
state logic_4 in, out	# Connection point declaration
{	# Start template body
when(event_on(in)) {	# Detect event on the input:
if (in == l4_0) \	# If input is "low"...
schedule_event(time, out, l4_1)	# ...turn output "high"
else if (in == l4_1) \	# If input is "high"...
schedule_event(time, out, l4_0)	# ...turn output "low"
else if (in == l4_x) \	# If input is "unknown"...
schedule_event(time, out, l4_x)	# ...turn output to "unknown"
}	# End of when statement
}	# End template body

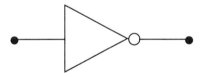

Figure 10-1. Digital inverter.

This template introduces several new concepts that are used for digital modeling:
- digital connection points (states)
- digital values (logic_4 states)
- **when** statements
- **event_on()** function
- **schedule_event()** function
- **time** as a simulator variable

The inverter template shown above does not have any arguments. Aside from this fact, the header line is quite similar to many that have been described in earlier chapters. The declaration of the connection points for this template uses the key terms **state logic_4** (4-state logic: 0, 1, X, Z).

Digital connection points

Thus far, all templates have used electrical connection points (pins). When an analog simulator encounters circuit nodes that are connections of electrical pins, it uses unit and pin definitions that specify current and voltage as through and across variables (see Chapter 5). These system variables can assume any value that is a real number.

A digital simulator is programmed to use state variables for nodes in a digital circuit. Consequently, for an AHDL model to describe digital behavior, it must be able to provide a different type of connection point—one that uses state variables. The MAST language provides such a connection point that is defined to assume only a restricted set of state values (such as 0, 1, or X).

Digital values

A state connection point cannot assume new values throughout a simulation analysis the way an analog connection point can. Changes in state values at a connection point must occur as scheduled events at discrete (specific) times. This is done using a **schedule_event()** statement in a template.

Although a state connection point can take a variable whose values are continuous or discrete, connection points for a *digital* template are declared as logic_4 states, which means that they use discrete values from a finite set. For example, a digital connection point may assume only the following logic_4 state values:

- **l4_0** (logic 0)
- **l4_1** (logic 1)
- **l4_x** (unknown)
- **l4_z** (inactive)

Digital Modeling

Note that the value of l4_x is active—it should be interpreted as "either l4_1 or l4_0 —but it is not known which." This differs from the meaning of l4_z, which means that none of the pins that have been driving a node are currently driving it (i.e., it is explicit notification that the bus is available).

State variables used as connection points must have their initial values scheduled as events (see Section 10.3). By contrast, internal (local) state variables with logic_4 units (i.e., those that are not connection points) may be assigned values (initialized) when they are declared in the template, the same way analog variables are. If they are declared but not initialized, it is up to the simulator to provide a default initialized value (usually the X-state value).

The when statement

The remainder of the inverter template associated with Figure 10-1 consists of a single **when** statement into which the event-driven functionality of the inverter is procedurally programmed. A **when** statement is the MAST construct that makes a template function as a digital model. (Chapter 11 will describe templates that contain both analog and digital functionality.) A **when** statement contains a conditional "trigger" that is a boolean expression governing evaluation of the rest of the statement. It appears in the following general form:

> **when**(*condition*) {
>   *statements*
>   }

where *condition* is one of the intrinsic functions or simulator variables listed below and *statements* can be either an expression or one or more calls to intrinsic scheduling functions **schedule_event**, **schedule_next_time** (see Chapter 11), or **deschedule**. When the *condition* is true, the *statements* are executed.

- **event_on()** is an intrinsic function that detects events coming off the event queue. When **event_on**(*state_variable*) becomes true (i.e., the event being "listened for" is now active), it will trigger the evaluation of the boolean condition of the **when** statement each time there is an event on the state variable.

- **threshold()** is an intrinsic function (described in Chapter 11) that detects changes in analog signals.
- **dc_init** is a simulator variable that becomes true at the beginning of a DC analysis or a DC transfer analysis.
- **dc_done** is a simulator variable that becomes true at the end of a DC analysis or a DC transfer analysis.
- **time_init** is a simulator variable that becomes true at the beginning of a transient analysis that is not re-started from a previous analysis.
- **tr_start** is a simulator variable that becomes true at the beginning of any transient analysis.

The schedule_event statement

To change the value of a state variable on a connection point so that it can be detected by other templates, the change in value must be placed in the event queue. This is done using the **schedule_event()** statement. An entry on the event queue contains three items, which are implemented as arguments to the **schedule_event()** statement:

- The time at which the event is to occur. This may be an expression (e.g., **time + 1m** in **inverter** template, which would create a 1 ms delay). The event cannot be in the past.
- The state variable that will use the value resulting from this event (e.g., **out** in inverter template).
- The value which the state variable will receive when the event is processed. This may be an expression.

Note that the variable **time** is available to the template as a *simulator variable* (see Appendix A). This is shown in the three **schedule_event** statements in the inverter template. An entry may be removed from the event queue using the **deschedule()** statement. Thus, the **when** statement section of the **inverter** is executed when an event is received on the **in** connection. Next, the if statement is executed—it basically checks the values of **in** and schedules the appropriate values to appear on the **out** connection.

10.3 Initialization

Analog simulation supports initialization through the DC analysis or by various techniques for acquiring steady-state initial points. In contrast, digital simulation, due to the nature of digital designs, leaves initialization up to the models. In fact, many stand-alone digital simulators do not even have a DC analysis or support the concept of an initial point.

In the case of DC analysis in a mixed-signal simulator, initialization of the digital system is considered to be complete when the event queue is empty (nothing left to process).[1] This means that in the digital simulator, during DC analysis, time is allowed to progress, whereas time in the analog simulator is kept at 0.

The MAST AHDL supports initialization through the following simulator variables:

- **dc_init**
- **dc_done**
- **time_init**
- **tr_start**

These variables may be used in the *condition* portion of a **when** statement (as listed in Section 10.2).

10.4 Digital Clock Template

Figure 10-2 shows an example of a digital clock, whose template is shown on the next page. This template introduces two new concepts:

- Local state variables
- DC and transient initialization

The clock template consists of three **when** statements. The first two are for initialization purposes. The last contains the time domain functionality of the template.

---

[1] In the case of an oscillation in the digital system, the analysis will terminate with an error when a large (but finite) number of events have been processed.

```
template clock out = f, duty
state logic_4 out                               # digital connection point
number  f=0,                                    # frequency argument
        duty=0.5                                # duty cycle argument
{
state nu tick                                   # internal counter
number  ton=0,                                  # clock on-time
        toff=0                                  # clock off-time
if (f > 0) {                                    # calculate off and on time
  ton = duty/f
  toff = 1/f - ton
  }

when (dc_init) {                                # DC initialization
  schedule_event(time, out, l4_0)
  }

when (time_init) {                              # transient initialization
  if (f > 0) schedule_event(time, tick, 1)      # (start clock ticking after delay)
}
when (event_on(tick)) {                         # clock pulse propagation
  if (driven(out)==l4_0) {

    if (ton > 0) {                              # turn clock on (set to 1)
      schedule_event(time, out, l4_1)
      schedule_event(time+ton, tick, 1)
      }
    }
  else {

    if (toff > 0) {                             # turn clock off (set to 0)
      schedule_event(time, out, l4_0)
      schedule_event(time+toff, tick, 1)
        }
      }
    }
  }
}
```

Digital Modeling

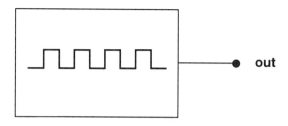

Figure 10-2. Digital clock symbol.

Header and local state variables

The digital clock template has only one connection point (**out**), since all other required characteristics can be specified via the input parameters (**f** and **duty**).

The only state variables contained in the **inverter** template in Section 10.2 are those associated with the connection points. It is also possible to declare a state as a local variable, such as **tick** in the **clock** template. While local states cannot be provided to a "parent" netlist (outside the template), they can be used in a local netlist (within the template). Local states can be initialized in the template without being placed in the event queue. Local states may also be on the left-hand-side of assignment statements in the body of **when** sections—if the state does not appear in an **event_on()** function. Other than these differences, local state variables are identical to state connection points.

Note that the units of **tick** are **nu** ("no units") instead of **logic_4**. That is because it is not being used to assign a logic state to a connection point. Here, **tick** serves as an internal counter to propagate clock pulses.

Initialization

The template uses two **when** statements for initialization:

- **when(dc_init)**—for DC analysis initialization. The **dc_init** simulator variable initializes the value at **out** to **l4_0** at the beginning of a operating point (DC) analysis.

- **when(time_init)**—for transient analysis initialization. The **time_init** simulator variable lets the template use simulation time (**time**) to synchronize clock timing and the **tick** counter so that clock cycles can be generated.

The **schedule_event** statements put **time** and **tick** in the event queue, so that the event-driven activity is recognized by the simulator. The third **when** statement is used to propagate the clock pulses, using the values of **ton** and **toff** (which are calculated from the argument values of **f** and **duty**) to determine pulse frequency and duty cycle, respectively.

10.5 Conflict Resolution and the Driven Function

Consider the MAST template for an AND gate as shown below.

```
template and in1 in2 out = td
state logic_4 in1, in2, out              # state connection points number td=0
{
state logic_4 out_state                  # internal state declaration
when (event_on(in1) | event_on(in2)) {   # beginning of when statement
    if ((in1==l4_1) & (in2==l4_1)) {     # AND logic
    out_state = l4_1
        }
    else out_state = l4_0
        if (driven(out)~=out_state) {    # update AND gate output
        schedule_event(time+td, out, out_state)
            }
        }                                # end of when statement
}
```

It would seem that the following:

```
if (driven(out)~=out_state) {
    schedule_event(time+td, out, out_state)
        }
```

Digital Modeling

could have been accomplished with:

```
if (out~=out_state) {
    schedule_event(time+td, out, out_state)
}
```

What was the purpose of the **driven** intrinsic function? The **driven** function is necessary because it is possible for a pin to be driven by more than one gate.

Figure 10-3 shows two logic gates driving the same node. The output of one gate is logic 1 and the output of the other is logic 0. What is the correct value at that node? It's ambiguous. It might be 1, it might be 0, and it might be something in between. This conflict is resolved by the conflict resolution mechanisms described below. The value of **out** *may be different* from the value of **driven(out)**. The **driven** function lets you proceed without having to worry about what is happening outside the template. The value of **out** reflects the value to which the output pin is finally set, whereas the value of **driven(out)** reflects the most recent value that the template has scheduled for the connection point, **out**.

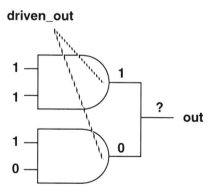

Figure 10-3. Conflict at a digital node.

Implementing conflict resolution

When two templates drive a single pin, there is potential for conflict. In this case a mechanism is needed to resolve the situation. The Saber simulator uses a routine specific to the logic_4 unit that enforces a truth table to resolve the conflict.

The body of Table 10-1 shows the resultant logic_4 value when a state along the left appears simultaneously with a state along the top. This is the truth table for conflict resolution for logic_4 unit states.

Table 10-1. Conflict resolution table for logic_4 states.

	l4_0	l4_1	l4_x	l4_z
l4_0	l4_0	l4_x	l4_x	l4_0
l4_1	l4_x	l4_1	l4_x	l4_1
l4_x	l4_x	l4_x	l4_x	l4_x
l4_z	l4_0	l4_1	l4_x	l4_z

For example, suppose that three different templates with logic_4 connections have their outputs connected to a single pin. Suppose further that the output of the first template is logic 0, that of the second is logic 1, and that of the third is logic Z (high impedance). The conflict resolution mechanism would then compare the first (l4_0) and second (l4_1) outputs to determine a resultant value according to the table—l4_x. That result would then be compared to the output of the third template (l4_z) to resolve the value of the node (l4_x).

Conclusions

When developing event-driven models, the thought process described in the earlier chapters basically still applies. However, those steps for the implementation of differential equations translate to the combinational or sequential logic functions required to implement digital behavior. Large-signal topologies and simultaneous equations give way to truth tables and event queues. In MAST, the differences are highlighted by changes in connection point types and by the replacement of values and equations sections with **when** sections.

CHAPTER 11

Modeling Mixed Analog-Digital Systems

In the previous chapter, event-driven constructs in MAST were described for modeling digital circuits and systems. However, if the system to be simulated is purely digital, then it could be more appropriate to use a digital simulator. The purpose for having digital capability in an AHDL is to allow the modeling and simulation of mixed analog-digital designs, or more generally, mixed continuous time/discrete time simulation. This includes some models that are purely digital, as described in Chapter 10. It also includes models that are themselves mixed analog-digital in nature. This chapter combines time domain analog modeling concepts with digital modeling to achieve mixed analog-digital models. Although this may seem complicated, it actually requires only a few new concepts, including:

- Using analog state variables
- Using the **threshold** intrinsic function to detect threshold crossings of analog signals
- Scheduling analog time steps with the **schedule_next_time** intrinsic function

Two examples are provided in this chapter. The first is that of a voltage comparator, which takes two analog inputs and produces a digital output. The second example is that of a digitally controlled, ideal switch, which uses a digital input to control the analog pins of the switch.

11.1 Modeling a Voltage Comparator

Figure 11-1 shows a comparator with two inputs and one output. The digital output value is based on a comparison of the two analog input values. If the voltage at **p** is larger than the input voltage at **m**, then the output is a logic 1 (**l4_1**); otherwise, the output is a logic 0 (**l4_0**).

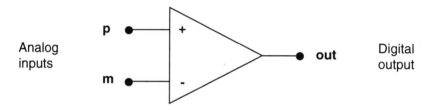

Figure 11-1. Voltage comparator.

The template for this simple voltage comparator model (**comparator**) is listed below.

```
template comparator p m out = td
electrical p, m                          # analog pins
state logic_4 out                        # digital connection
number td=0                              # argument
{
state nu before, after                   # local variables
when (threshold(v(p), v(m), before, after)) {
    if (after > 0)      schedule_event(time+td, out, l4_1)
    else                schedule_event(time+td, out, l4_0)
    }
when (dc_init) {                         # DC initialization
    schedule_event(time, out, l4_0)
    }
}
```

Modeling Mixed Analog-Digital Systems **135**

The header declarations show that this voltage comparator template has two analog (**electrical**) inputs and one digital (**state logic_4**) output. This indicates that it is a *mixed* analog-digital template. Because the digital output is a discontinuous state type, as introduced in Chapter 10, the output of the **comparator** template could be used as an input to the **and** template of that chapter. The time delay argument (**td**) is also similar to the one introduced for the **and** template; it specifies a delay between the time the input voltage changes polarity to the time the output changes state.

This model uses a **when** statement to compare the analog voltages at the two input nodes to produce a digital output of logic 1 if the voltage at **p** is greater than the voltage at **m**, and logic 0 if otherwise.

The local variables for this template are **before** and **after**. These variables determine which direction the input voltages are going after crossing one another. They are simultaneously **state** variables and analog variables. As analog variables, they are not limited to a discrete set of *values*—they can take on any real-numbered value. As **state** variables, they can change their values discontinuously in *time*. This type of variable is known as an *event-driven analog* (or analog state) variable.

The units of **before** and **after** are declared as **nu**, which stands for no units. In general, the units for analog **state** variables can be any of those in **units.sin** (see Appendix A)—the same way a **val** or a **var** can assume these units.

The **before** and **after** variables must be of type **state** because their values are set in a **when** statement. Any variable that is set in a **when** statement (either by an assignment statement or by **schedule_event**) must be of type **state**.

In this template, there are two **when** statements. The first detects the crossing of the two input voltages. The second is for DC initialization. Note that in the **clock** template of Chapter 10, the **when** statements for initialization appeared before the time domain behavioral **when** statement. In this template, the initialization **when** statement comes last. The order of **when** statements is arbitrary.

However, the statements *within* the **when** statements are executed in order. The first **when** statement from the comparator template detects the crossing of the two input voltages. It can be interpreted to mean that when the voltage at **p** crosses the voltage at **m**, set the values of **before** and **after** to their appropriate values (as described below), and execute the **if-else** statements that follow.

The **threshold** intrinsic function is simply stated but powerful; it pinpoints the exact time when two changing quantities become equal and provides information about the past and the future of those quantities. In this instance, it monitors the voltages at connections **p** and **m**, then "triggers" an event during the simulation at the precise time that the voltages become equal to each other. This functionality is independent of the simulator time steps.

This threshold checking sets the values of **before** and **after** according to the following (see Table 11-1).

- If the first argument of the **threshold** statement, v(p), is crossing the second argument, v(m), from negative to positive, then **before** is set to -1 and **after** to +1.
- If the first argument of the **threshold** statement is crossing the second from positive to negative, then **before** is set to +1 and **after** to -1.
- If v(p) and v(m) remain equal after becoming equal, then **after** is set to 0.

Table 11-1. Threshold crossing combinations.

before	after	meaning
-1	0	v(p) rose from below v(m) to equal v(m)
+1	0	v(p) fell from above v(m) to equal v(m)
0	-1	v(p) fell from equal to v(m) to below v(m)
0	+1	v(p) rose from equal to v(m) above v(m)
+1	-1	v(p) fell from above v(m) to below v(m)
-1	+1	v(p) rose from below v(m) to above v(m)
-1	-1	v(p) rose to equal v(m) then fell below v(m)
+1	+1	v(p) fell to equal v(m) then rose above v(m)

Now that there is a way to find the voltage crossing point, completion of the comparator function is simple. The value of the **after** variable (shown in Table 11-1) is used in the if-else statement to determine whether v(p) is crossing positive or negative. If the crossing is positive, then **out**, the output pin, is scheduled to be l4_1 (after time delay **td**). Otherwise, **out** is scheduled to be l4_0.

Thus, the **threshold** function and the **if-else** comparison of the first **when** statement implement the comparator model for a transient analysis. However, for a DC analysis there are further considerations. The purpose of the DC analysis is to determine the quiescent values of all of the system's internal variables and states. The interactions between these quantities can be complex. The algorithm used by the Saber simulator for a DC analysis is specifically designed to manage these complexities.

One of the major problems that arises from these interactions is that it is not always possible to find a stable solution. A good example of a system that has no single stable operating point is a ring oscillator. When the DC algorithm detects an oscillation, it arbitrarily selects a state at a node and allows the unsatisfied scheduled events to occur during the transient simulation.

To understand these state interactions, it is necessary to look at how the DC analysis of the simulator pertains both to the scheduling and detection of events and to the use of the **threshold** function. A simplified description of the DC analysis[1] for the Saber simulator is as follows:

1. The simulator sets all analog independent variables to *zero*, all internal event-driven states (analog or digital) to their initialized values (typically this is the default value assigned upon declaration or a passed-in value if the state is an argument to the template), all analog state connection points to **undef**, and all digital connection points to l4_x.

2. It executes **when (dc_init)** statements in all templates, propagating all events according to dependencies.

3. The effects of the analog subsystem on the discrete subsystem are found by observing all **threshold** conditions, as analog signals go from *zero* to their final DC values. For

---

[1] This analysis technique has been patented (No. 4,985,860).

any threshold conditions that become satisfied, the *statements* portion of the corresponding **when** statement is executed. The simulator again propagates all events within the discrete subsystem according to dependencies.

4. Repeatedly, the simulator finds the solution of the analog subsystem solution and propagates all events within the discrete subsystem, until the system stabilizes.

Consider the DC initialization provided by the second **when** statement in the **comparator** template:

```
when (dc_init) {
    schedule_event(time, out, l4_0)
}
```

If this were not included in this template, the output could potentially be incorrect at the end of a DC analysis. This is demonstrated by the circuit shown in Figure 11-2.

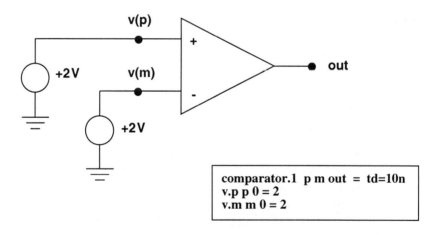

Figure 11-2. Comparator DC analysis.

If the **when (dc_init)** statement were omitted and **v(p)** were 2 V and **v(m)** were 2 V, then the DC algorithm would proceed as follows:

1. Start with 0 V at **p** and **m** and with the state of the output pin **out** at its initial value of l4_x (logic X—unknown state).

2. Execute all **when (dc_init)** statements. (Assuming that the **when (dc_init)** statement is absent, there is no action here— **out** remains at l4_x.)

3. As **v(m)** and **v(p)** increase together from 0 (the initial values) to 2 volts (the source voltage), *the threshold is NOT crossed, and the output remains at* l4_x. This is the value at **out** at the end of the DC analysis, which is erroneous. The expected output would be logic 0 (l4_0), because **v(m)** is equal to **v(p)**. However, by this algorithm, the output pin **out** was set to logic X and would remain there.

Adding the second **when** statement for DC initialization alleviates this problem by initializing the output pin **out** to logic 0.

If **v(p)** is set to a DC value greater than **v(m)**, then the threshold will take care of producing the correct DC value for the output, because the value of **v(p)** would cross the value of **v(m)** in step 3 of the algorithm above.

NOTE

A **when (threshold())** *statement should be accompanied by a* **when (dc_init)** *statement that establishes the appropriate conditions, assuming that all node voltages and system variables are 0. If the threshold is crossed, then the conditions established by the* **when (dc_init)** *statement are overridden and produce the expected result.*

11.2 A Digitally-Controlled, Ideal Switch

This example shows an ideal, single-pole, single-throw switch (SPST) controlled by a digital input. It has two electrical pins and one digital connection as shown in Figure 11-3.

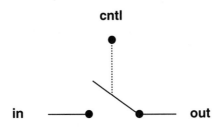

Figure 11-3. Ideal switch

The template for this ideal switch (sw) is listed below.

```
template sw p m cntl = ron, roff
electrical p, m                          # analog pins
state logic_4 cntl                       # digital connection
number ron=1, roff=1meg
{
branch cur=i(p->m), vlt=v(p,m)
state r res = roff                       # internal state variable
when(dc_init) {                          # DC initialization
    res = roff
    }
when (event_on(cntl)) {                  # switch control
    if (cntl == l4_1)      res = ron
    else                   res = roff
    schedule_next_time(time)
    }
# template equation for analog switch
cur = vlt/res
}
```

This model of an ideal SPST switch demonstrates another aspect of the mixing of analog and digital concepts. In this model, a digital event comes to the template through the cntl input pin. If the cntl pin is logic 1, then the switch is closed. If the cntl pin is logic

0, logic X, or logic Z, then the switch is open. The resistance between the electrical pins **p** and **m** is equal to **roff** when the switch is open; the resistance equals **ron** when the switch is closed. The resistance changes instantaneously when the event occurs on the **cntl** pin.

The header shows that this ideal switch template has two electrical pins (**p, m**), one digital connection point (**cntl**), and arguments for on and off resistances (**ron, roff**). Note that the connection points are both analog pins (**electrical**) and digital states (**state logic_4**). The template arguments, **ron** and **roff**, are initialized to default values of $1\,\Omega$ and $1\,M\Omega$, respectively. Thus, it is not necessary to assign values to **ron** and **roff** to use this template.

Local declarations include the branch current (**cur**) and voltage (**vlt**) from **p** to **m** along with the analog **state** variable **res**. The variable named **res** has the units of resistance. This variable contains the present value of resistance, whose value changes discontinuously. It can be assigned the value of a passed-in argument, as shown. It cannot be assigned anywhere except in a **when** statement, but it can be referenced anywhere that **state** references are permitted (e.g., used in the right-hand side of an assignment statement). In this example, the value of **res** is referenced in the template equation. The template relies on detecting an event coming in during the DC analysis. To ensure that **res** has a value when the transient analysis begins, a default value can (and should) be assigned as follows:

    state r res=roff

The **when (dc_init)** statement initializes the value of **res** to **roff**. Note that the variable **res** must be of type **state**, because its value is set in a **when** statement. This is true for any variable whose value is set in a **when** statement, either by assignment or by a **schedule_event** function call.

Once again, a **when** statement is used to implement the digital behavior of the model. The **event_on** function is used in the **when** statement to determine at what time the state of the switch changes from open to closed, or vice-versa. When an event occurs at **cntl**, it is a simple matter to check to see whether it has changed to logic 1 (**l4_1**) or logic 0 (**l4_0**). If it is **l4_1**, then the local **state** variable for resistance, **res**, is set to the argument value for the on resistance, **ron**. Otherwise, the value of **res** is set to **roff**.

The template equation for the **sw** template is straightforward. The current from **p** to **m** is the result of dividing the voltage across the nodes to which they are connected by **res**, the present resistance value. This resistance happens to be a state variable whose value is determined by events at the control pin **cntl**. As a result, the value of **res** is changing discontinuously in time. This does not generally cause a problem, because **schedule_next_time** restarts the integration algorithm with each event, using the value of **res** from the previous event as an initial point. However, it is best practice to provide continuity.

Analog time steps

Normally, analog time steps are determined by the analog simulation time-step algorithm, which uses variable time steps and an integration algorithm to ensure that the time-steps are as big as possible, without causing too much error in the solution at any given time.

It is important to note that the analog subsystem time points are completely independent of the discrete subsystem time (event) points. The **threshold** function lets the analog subsystem influence events in the discrete subsystem. On the other hand, the **schedule_next_time(time)** function forces an analog time step when a transient simulation reaches **time**.

Therefore, it is important to include the **schedule_next_time** in the **when** statement. The fact that an event occurred on the **cntl** pin does not mean that there will be an analog time step when **res** changes value. Instead, the template must force an analog time step to occur at that time by scheduling it with **schedule_next_time**. This lets the new value of **res** affect the analog network at the right point in time.

Note that there are some subtle usage characteristics of the **schedule_next_time** function. Its objective is to force a time step in the analog simulation. However, it is important to note that once scheduled, it *stays in effect* until either it is descheduled or the time point has been reached.

Example circuit

Figure 11-4 shows an example demonstrating the use of the **sw** template with a voltage source and a resistor. The **when** statements in the netlist provide a digital controlling stimulus for the switch at node **gt**. Note how the language allows direct inclusion of these statements in the netlist (i.e., an additional model was not required to provide the digital input at **gt**).

```
sw.1 in mid gt = ron = 1, roff = 47k
r.1 mid 0 = 47k
v.in in 0 = 5
when (dc_init) {
    schedule_event(time, gt, l4_0)
}
when (time_init) {
    schedule_event(1u, gt, l4_1)
    schedule_event(2u, gt, l4_0)
}
```

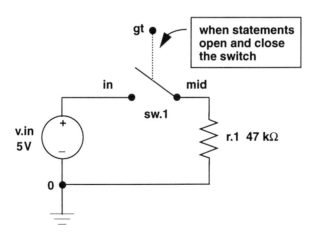

Figure 11-4. *Digitally-controlled switch in an analog circuit.*

Conclusions

As shown in Chapter 10, the MAST AHDL is not limited to providing models only for continuous time (analog) simulation. Moreover, MAST can model both digital and analog characteristics—even within the same template. Consequently, templates for analog-to-digital [1] and digital-to-analog converters can be easily created. Because of this, MAST becomes particularly well-suited for interacting with a digital simulator in a mixed-signal simulation environment [2].

11.3 References

1. G. Ruan, "A Behavioral Model of A/D Converters Using a Mixed-Mode Simulator," *IEEE J. of Solid-State Circuits*, Vol. 26, No. 3, pp. 283-290, Mar. 1991.

2. *Mixed Signal Design Simulation*, Pub. MB-0194, Analogy, Inc., 1994.

CHAPTER 12

Advanced AHDL Capabilities

One measure of a good modeling language is that it allows users to express their problems naturally. Previous chapters demonstrate the "through and across variable" approach for analog networks and the "event-driven" approach for digital networks. These approaches are quite common in representing electrical circuits. However, for an AHDL to be truly helpful in top-down design as described in Chapter 4, it must also be able to represent system-level models. This includes the capability to model functional blocks without becoming burdened with conservation laws and electrical nomenclature that are unnecessary at this level.

This chapter introduces an approach for representing systems. This is accomplished by connecting functional blocks by using **var** and **ref** connections, which are abbreviations for variable and reference. A **var** is an independent system variable for which the simulator must solve, and a **ref** is a reference to a variable declared as a **var** in another template.

This modeling approach assumes that each functional block has distinct inputs and outputs and that the outputs are a function of the inputs. Because the inputs affect the outputs, but not vice versa, there is a notion of *direction* to the flow of information through the functional blocks.

For example, consider a functional block that takes two inputs and produces their sum as the output. The flow of information is from the summing inputs to the output.

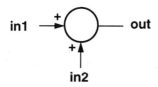

Figure 12-1. Functional summing block.

Because there is no physical quantity to be conserved, it isn't necessary to create an equivalent circuit network to model this block. The techniques described in this chapter show how to model it directly. In control system modeling, much of the information flows through these types of functional blocks. This chapter uses control system examples to illustrate information flow through functional blocks. However, the concepts also apply to other areas such as the z-domain.

Another important AHDL topic is that of statistical modeling. Section 12.3 describes how MAST incorporates statistical distributions and correlations of parameters in models.

Section 12.4 covers noise modeling. While noise simulations are utilized to a lesser extent that other simulations, it is an important topic nonetheless, because noise models are incorrect in many traditional model implementations.

12.1 Control System Modeling (s-Domain)

Control system models are sometimes referred to as *signal flow* or *data flow* models. This chapter illustrates how to model control system blocks using the following:

- **input** and **output** connection points (which are similar to **ref** and **var** connections)
- s-domain modeling
- ideal delay modeling
- z-domain modeling

Using **var** and **ref** connections provides certain advantages for describing the flow of information among various functional blocks. In modeling control system blocks, the output of one template (a **var**) is provided directly as the input (a **ref**) to a second template. These connection points do not require compliance with conservation laws (such as KCL), which means there does not need to be a reference node in the circuit. As a convenience, the MAST language provides two connection point types that provide the same functionality as **var** and **ref**, called **output** and **input**, respectively. They are declared as connection points in the same way as **var** and **ref** are declared.

Control system concepts can be illustrated by developing templates and combining them hierarchically (as covered in Chapter 6) into a system model such as that shown in Figure 12-2. The input to such a system comes from a "source" template, which is similar to voltage source templates described in previous chapters, but which has only a single connection (similar to the digital clock of Chapter 10), declared as an **output** type. The other templates in this design are developed using **output** and **input** connection points, s-domain modeling, and ideal delay modeling.

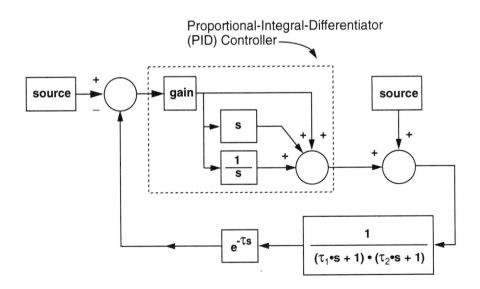

Figure 12-2. Control system diagram.

Creating basic control system templates

This section describes the basic approach for writing control system templates using two examples:

- A simple source (**dcsrc**)—although not used in the example circuit shown in Figure 12-2, this example illustrates the similarity to the simple voltage source shown in Chapter 7.
- A two-input summer (**sum2**)—this example shows how to write a control system template whose output is the sum of its two inputs.

<u>*Simple source template*</u>

The DC source template (**dcsrc**) listed below is the simplest example of a control system template, because it has no inputs and only one output. This template has one connection point, **p**, which is declared as an **output**.

```
template dcsrc p = vs
output nu p
number vs
{
p = vs
}
```

Compare this template to the simple *voltage* source template (**vsource**) provided in Chapter 7:

```
template vsource p m = vs
electrical p, m
number vs
{
branch i=i(p->m), v=v(p,m)
v = vs
}
```

Note the following differences between these two templates:

1. The **dcsrc** template has only one pin (**p**), which is unitless and is an **output** type. The **vsource** template has two **electrical** pins. The **vsource** template needs at least two pins (**p, m**)

Advanced AHDL Capabilities **149**

because it uses an across variable (voltage), which requires a circuit reference node. The **dcsrc** template needs only one pin because it is a simple number source with no units (**nu**).

2. The **dcsrc** template does not model conservation of any physical quantity; its output is a unitless variable that requires no "path" to flow through or to be measured across. The **vsource** template models the conservation of electrical quantities (current and voltage), and declares its pins (**p, m**) to be a branch for these quantities.

3. In the **dcsrc** template, the variable at **p** is provided as a **var**, which means it can be used directly in the template equation (the value of a **var** is always referenced directly). It can also be used as a connection to other models. In the **vsource** template, the through and across variables (current and voltage) at the output pins are referenced in the template equation by means of the branch notation.

Two-input summer template

This section shows how to create a template that can take input stimuli, operate on them, and produce an output. The example is a two-input summer, called **sum2**.

```
template sum2 in1 in2 out = k1, k2
input nu in1, in2
output nu out
number k1=1, k2=1
{
out = k1*in1 + k2*in2
}
```

Note that the two inputs (**in1** and **in2**) are declared as **input** connections with no units (**nu**). The output (**out**) is declared as an **output** connection with no units.

The output is defined as the sum of products, each the product of an input and its corresponding constant. Typically, **k1** and **k2** are either 1 or -1, depending on the desired function. If **k1** is 1 then **in1** is added, whereas if **k1** is -1, then **in1** is subtracted. A similar statement applies to **k2** and **in2**. In general, **k1** and **k2** can be any desired real numbers.

Example

Figure 12-3 shows how two instances of the **dcsrc** template can be used as inputs to the **sum2** template. Given the following netlist and argument values (**sum2** gain arguments are left at their default values of 1):

    dcsrc.1 in1 = vs=5
    dcsrc.2 in2 = vs=7
    sum2.1 in1 in2 out1

the completion of a DC or transient analysis using this system would produce the following values for the signals:

    in1 5
    in2 7
    out1 12

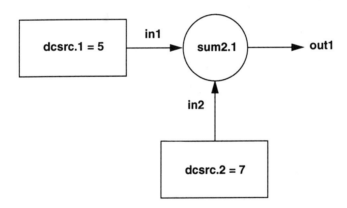

Figure 12-3. Connection diagram for **sum2** and **dcsrc** templates.

Laplace domain modeling

In control systems modeling, the system description is often available in s-domain form. Figure 12-2 shows several blocks where the block transfer function is defined as an s-domain function. The letter "s" is standard notation for the complex frequency quantity $\sigma+j\omega$, where σ is the real part and ω is the imaginary part (j is $\sqrt{-1}$ and ω is frequency in radians/second). The "s" quantity is used when solving time-domain differential equations using

Laplace transforms. The Laplace transforms and inverse Laplace transforms make it possible to perform calculations in either the time-domain or the s-domain.

The advantage of using the s-domain is that complicated operations in the time-domain are made easier in the s-domain. For example, differentiation in the time-domain is multiplication by s in the s-domain. Similarly, integration in the time-domain is simply division by s in the s-domain. Initial conditions must be established for integration, which is explained for the integrator template later in this chapter.

The MAST modeling language does not directly implement s-domain modeling. However, a simple rule is given below for converting s-domain expressions into time-domain expressions and simplifying the operation: whenever a quantity in the s-domain expression is multiplied by s, substitute the multiplication operation with a differentiation operation, using the MAST differentiation operator, **d_by_dt**.

This process is illustrated below, using the following examples:

- a differentiator (using multiplication by s)
- an integrator (using division by s)
- a two-pole transfer function (using an s-domain polynomial in the denominator)

Note that integration can be performed without the explicit need for an integration operator—the differentiation operator (**d_by_dt**) is sufficient.

Differentiator template

Figure 12-4 shows a diagram for a differentiator. This illustrates the equivalence of multiplication by s in the s-domain to differentiation in the time-domain. The output is the time derivative of the input multiplied by a constant (parameter **tau**).

The MAST template for this differentiator (**deriv**) is shown below. Note that the constant multiplier, **tau**, must be inside the **d_by_dt** operator in the template equation. The simulator finds the value of **out**, such that **out** is equal to the time derivative of the product of **tau** and **in**.

```
template deriv in out = tau
input nu in
output nu out
number tau
{
out = d_by_dt(tau*in)
}
```

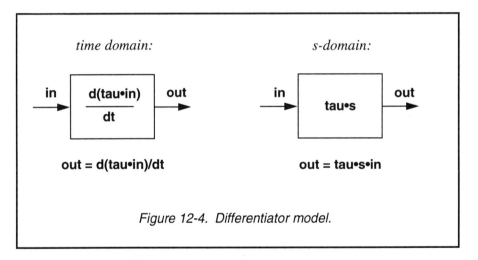

Figure 12-4. Differentiator model.

Integrator template

As mentioned earlier, integration in the time domain is equivalent to division by s in the s-domain, which could be expressed as:

$$out = \frac{in}{tau \cdot s} \qquad (12\text{-}1)$$

To implement this as a valid template equation, the following steps need to be performed:

1. Revise the statement so that s is not in the denominator. To accomplish this, simply multiply both sides by **tau*s**, which yields:

   tau*s*out = in

Advanced AHDL Capabilities

2. Replace the s operator with the MAST **d_by_dt** operator, and move **tau** inside the **d_by_dt** operator. This sets the time derivative of the output multiplied by a constant (**tau**, the template argument) equal to the input.

   d_by_dt(out*tau) = in

3. Insert the keyword **make** at the beginning of the statement:

   make d_by_dt(out*tau) = in

This is required because the lefthand side of Step 2 above is not a simple output, but an expression, which means this relation cannot be expressed directly. Statements such as this must be prefixed with the reserved word **make** so it will be recognized as a template equation.

As described in Chapter 5, the reserved word **make** is required anytime the output variable is embedded in an expression. It can also be used (although it is not required) if the lefthand side is an output (such as for the **deriv** template).

The MAST template for an integrator (**intgr1**) is listed below and illustrated in Figure 12-5.

```
template intgr1 in out = tau
input nu in
output nu out
number tau
{
make d_by_dt(out*tau) = in          # solve for the output variable
}
```

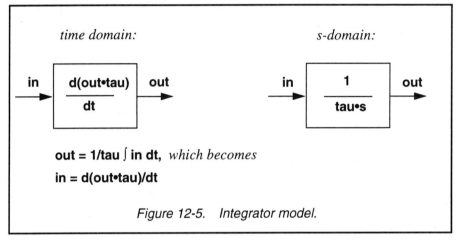

Figure 12-5. Integrator model.

This integrator template operates as expected for transient and small signal AC analyses. However, it produces a singularity in the DC analysis. This is because the equation, expressed in integral form, is as follows:

$$out = 1/\tau \int in\ dt + C \tag{12-2}$$

where C is a constant of integration. The MAST implementation effectively differentiates both sides of the integral equation. The constant of integration is therefore lost.

One way to handle the constant of integration is to change the integral from an indefinite integral to a definite integral, by specifying the initial condition for the **out** variable. The following template, **intgr2**, shows this modification to **intgr1** that includes a user-specifiable initial value (**init**) for the output.

```
template intgr2 in out = tau, init
input nu in
output nu out
number tau, init=undef
{
control_section{
    initial_condition(out,init)
    }
make d_by_dt(out*tau) = in
}
```

By definition, the DC solution to a system is the solution to the system when all time derivatives are set to 0. DC initialization is handled by using an **initial_condition** statement in the control section. Here, this is implemented in two steps:

1. Declare an argument (**init**) that allows the user to specify a value to be assigned to **out** at DC. Here it appears on the same line as **tau** and has been given a default of **undef**.

2. Create a control section in the template, and insert an **initial_condition** statement within it.

This technique allows one to specify the **init** parameter of the **intgr2** template as the output of the template during the DC analysis. Note that a value must be specified for **init** when the template is used because the default value of **undef** has no functional effect through the **initial_condition** statement. If the analysis is in the time-domain or frequency domain, the template equation (which is identical to the original template equation for **intgr1**) is then in effect.

Implementing two poles

To obtain an s-domain transfer function of order greater than one, the **d_by_dt** operator must be used each time the s operator is used as a factor. For each additional differentiation performed, an additional system variable (**var**) must be introduced. This is illustrated below for a two-pole transfer function.

In the s-domain, the two-pole transfer function (with poles represented by **tau1** and **tau2**) is expressed as:

$$\frac{out}{in} = \frac{1}{(tau1 \cdot s + 1) \cdot (tau2 \cdot s + 1)} \qquad (12\text{-}3)$$

Clearing the fractions by multiplication of both sides by the denominators and expansion produces:

$$tau1 \cdot tau2 \cdot s^2 \cdot out + (tau1 + tau2) \cdot s \cdot out + out = in \qquad (12\text{-}4)$$

At this point, the **d_by_dt** operator is substituted wherever the s operator appears. When **s**2** is encountered, a new **var** must be introduced for the result of the first differentiation (**doutdt**), then differentiated again. The template for this two-pole transfer function (**twopole**) is shown below.

```
template twopole in out = tau1, tau2
input nu in
output nu out
number    tau1,      # time constant of first pole
          tau2       # time constant of second pole
{
var nu doutdt        # var for finding first derivative
equations {
    doutdt: doutdt = d_by_dt(out)
    out: in = d_by_dt(tau1*tau2*doutdt) + (tau1+tau2)*doutdt + out
    }
}
```

Note the explicit usage of **var** as opposed to **output**, since **doutdt** is not a connection point. It is also noteworthy to point out that the **equations** section header is required once an intermediate variable (**doutdt**) has been introduced.

Ideal Delay

The s-domain transfer function for an ideal delay is:

$$out/in = e^{-s \cdot (tdelay)} \qquad (12\text{-}5)$$

where the output is identical to the input, but delayed by an amount of time represented as tdelay. This transfer function can be implemented in MAST using the **delay** function, rather than replacing the quantity s with the operator **d_by_dt**. The **delay** function makes it a simple matter to write a template for an ideal delay (**dlay**).

The quantity s can be replaced by the **d_by_dt** operator only in expressions involving linear combinations of s, which is not the case above.

```
template dlay in out = tdelay
input nu in
output nu out
number tdelay = 0                    # time delay from input to output
{
out = delay(in,tdelay)
}
```

12.2 Sampled Data System Modeling (z-Domain)

The MAST language has the capability to model the behavior of sampled data systems (SDS) using the mathematical properties of event-driven analog values in the z-domain [1, 2, 3]. It is possible to develop models for algebraic, integral-differential, and rational polynomial functions similar to those done in the s-domain. One significant difference between s-domain and z-domain modeling is the connection point type. Sampled data models use *analog states* as connection points rather than **vars** (output) and **refs** (input). This is because these templates model the effects of sampling continuous analog signals at discrete points in time.

Sampled data system models are based on the underlying principle of describing a function at discrete points in time, rather than continuously. A discrete time function, in turn, employs the concept of *sampling* input and/or output values, which can be thought of as repeatedly closing and opening a switch. Thus, a sample is taken (i.e., the switch is closed for an instant of time) every T seconds. This value of T is the sampling period in seconds. It is equal to 1/f, where f is the sampling frequency in Hz.

If the input to an ideal sampler is represented as r(t), its output is r*(t), and the present sample time is nT [1]. Thus, the present value of r*(t) is r(nT). It is possible to express any given value of r*(t) in terms of the impulse function (δ) as:

$$r^*(t) = r(nT)\delta(t - nT) \qquad (12\text{-}6)$$

Assuming that multiple samples are taken over the range of the discrete time function, then a series of samples of a given signal is expressed as:

$$r^*(t) = \sum_{k=0}^{\infty} rkT\delta(t - kT) \qquad (12\text{-}7)$$

From Eq. (12-7) for the sampled output, r*(t), it is possible to use the Laplace transform on signals (for t > 0) such that:

$$\mathcal{L}\{r^*(t)\} = \sum_{k=0}^{\infty} r(kT) e^{-kTs} \qquad (12\text{-}8)$$

This results in an infinite series that uses multiples of e^{sT}, allowing the following definition ($z = e^{sT}$), establishing a conformal mapping from the s-plane to the z-plane, from which follows the *z-transform*.

$$\mathcal{Z}\{r(t)\} = \mathcal{Z}\{r^*(t)\} = \sum_{k=0}^{\infty} r(kT) z^{-k} \qquad (12\text{-}9)$$

In general, the z-transform of a function f(t) is expressed as the following (where f[k] stands for the value of f(t) when t = kT).

$$\mathcal{Z}\{f(t)\} = F(z) = \sum_{k=0}^{\infty} f[k] z^{-k} \qquad (12\text{-}10)$$

Advanced AHDL Capabilities

As with the Laplace transform, it is the output of the system in the original domain (time) that is of interest; therefore, an inverse z-transform is also required to obtain output f(t) from F(z).

A summing block in the z-domain

The **sumz** template below shows how to implement a two-input summing block in the z-domain.

```
template sumz zin1 zin2 zout = gain, delay
state nu zin1, zin2, zout              # input and output states
number    gain = 1,                    # transfer gain
          delay = 1n                   # delay of block
{
state nu znew, zold=0                  # internal states of zout
when(event_on(zin1) | event_on(zin2)) {
    znew = gain*(zin1 + zin2)          # determine the new output state
    if(znew ~= zold) {
        schedule_event(time+delay, zout, znew)
        zold = znew
        }
    }
}
```

12.3 Statistical Modeling

Previous chapters describe techniques for modeling a particular component or design. The object being modeled is described with equations and accompanying coefficients (model parameters). In those chapters, the values of model parameters are assumed to be constants that characterize the object. This is a good assumption when the model represents a single sample of a component or circuit. However, another sample of the same component or circuit might be better characterized by a slightly different set of model parameter values, whose variations are due to the tolerances of the components.

This chapter introduces the notion that a parameter may be best described as a collection of possible values, with each value having its own likelihood of occurring. The name *statistical mod-*

eling describes the process of varying model parameters in a precise, yet random, way.

The following statistical modeling constructs allow the definition of a model with built-in variability:

- Probability density functions
- Cumulative density functions
- Correlated distributions
- Randomized variables

A model containing these constructs can be used with a simulator running what is known as a Monte Carlo analysis. This type of analysis performs a series of simulations, where each simulation uses a new set of parameter values according to the statistical distributions specified in the model.

Probability density function (PDF)

It is not possible to specify the probability distribution of a continuous random variable by listing all possible values of the variable and their corresponding probabilities. However, the probability distribution of a continuous random variable can be represented by a probability density function (PDF), which is the primary method for describing parameter variations.

A PDF is a continuous function of an independent variable (such as x) such that, for real numbers a and b, the probability that a random value of x will be between a and b is the area (density) under the PDF curve between a and b. MAST provides two predefined probability density functions that can be specified for any template parameter—**normal** and **uniform**.

> **normal**   The density function represented by the bell-shaped curve (sometimes referred to as "Gaussian") of Figure 12-6a. The x-axis shows all possible values of a parameter (continuous for all parameter values between $-\infty$ and ∞) and the y-axis shows the probability density for each possible parameter value. The total probability for all x-axis values (i.e., the area under the curve) is one.

This distribution represents degrees of deviation from the central tendency—the occurrence of the maximum density value for the parameter value at the middle of the curve. Note that lower probability densities occur for parameter values that move away from the middle in either direction. The *mean* or *average* of a normal distribution specifies the peak x-axis value in the middle; the *standard deviation* (σ) specifies a distance from the mean along the x-axis (in either direction) that will contain a known area of the curve.

uniform The density function represented by the rectangular curve of Figure 12-6b, where the x-axis shows all possible values of a parameter within an interval defined by two specific real values. The y-axis shows a uniform (constant) probability density for each possible parameter value within that interval. The total probability for all x-axis values (i.e., the area under the curve) is one.

It is also possible to create a piecewise linear PDF (such as the triangular curve shown in Figure 12-6c), but it requires more complicated constructs.

The method for applying one of these statistical distributions to a parameter value is to specify the name of the distribution, the nominal value (mean), an upper tolerance value, and a lower tolerance value. Note that, for a normal distribution, the upper value is three standard deviations (3σ) above the mean, and the lower value is three standard deviations (-3σ) below the mean.

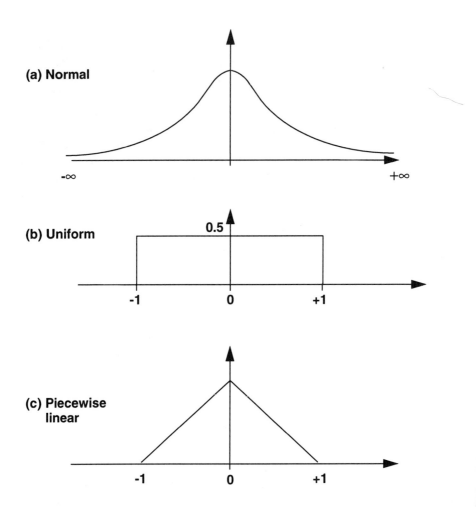

Figure 12-6. Normal, uniform, and piecewise linear PDFs.

For example, instead of specifying a fixed value for the nominal resistance argument (**res**) of the **resistor** template from Chapter 7, a normal distribution for **res** in a netlist entry could be specified as follows:

r.r1 in 0 = res=normal(1k, 910, 1090)

This would provide a normal distribution for values of **res** in **r.r1** with a mean of $1\,k\Omega$ and a standard deviation of $30\,\Omega$ (note that 910 and 1090 specify 3σ and -3σ values).

Instead of specifying separate values for upper and lower tolerances, a tolerance value that is a multiplier of the nominal can be specified. This multiplier automatically provides ±3σ values, so that the preceding example could also be expressed as:

    r.r1 in 0 = res=normal(1k, 0.09)

This same construct is available for the uniform distribution, but the upper and lower limits define absolute largest and smallest values instead of ±3σ points.

Cumulative density function (CDF)

The probability density function (PDF) is a common way of specifying the statistical variations of a design parameter. However, sometimes it is more convenient to specify a cumulative density function (CDF). Both the PDF and the corresponding CDF describe the same distribution, but they do so in slightly different ways. The CDF, like the PDF, is a function of x, where x ranges from $-\infty$ to ∞. The value of a CDF at x is the integral of the PDF, evaluated between $-\infty$ and x.

In other words, the CDF at any point x is the probability that a sample from the distribution has a value less than x. Obviously, at x=∞ the value of the CDF function must equal 1, because the probability that a sample from any distribution will be less than ∞ is 1. Correspondingly, at x=$-\infty$, the CDF function must equal 0, because the probability that a sample from any distribution will be less than $-\infty$ is 0. Figure 12-7 shows an example of a uniform PDF and its corresponding CDF.

As as result, the only intrinsic CDFs provided in MAST correspond to piecewise linear PDFs.

Correlation

Sometimes it is desirable to model two or more quantities that tend to vary together. For example, two resistors may be manufactured on an integrated circuit. The resistor values may vary a great deal from wafer to wafer, or even from die to die. However, the resistors on the same die may tend to vary together. That is, if one resistor is at the high end of its range, others from that die tend to be at the high end of their ranges as well. When this occurs, the resistor values are said to be *correlated*. Correlation occurs in numerous actual applications. This section explains how this kind of variation can be modeled with constructs already described.

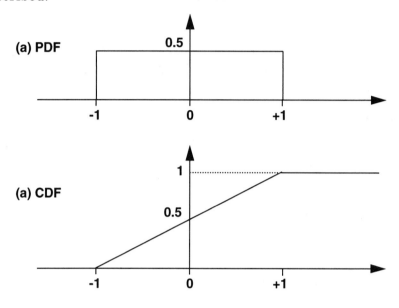

Figure 12-7.   Uniform PDF and corresponding CDF.

For example, assume that two resistor values are uniformly distributed with a tolerance of 10%, and that they correlate with each other within 0.5%.

The following netlist implements these relationships:

```
number common=uniform(1,0.1)
r.1    in mid  = normal(common*470k, 0.005)
r.2    mid 0   = normal(common*100k, 0.005)
```

The first line declares an arbitrarily named variable called **common** and uses the **uniform** function to assign an initial value from a uniform distribution. This distribution defines a uniform probability to all values between 0.9 and 1.1 (±10% tolerance around the nominal value of 1). The resistor netlist entries (**r.1, r.2**)

Advanced AHDL Capabilities 165

use the value of **common** as a multiplier to provide the desired correlation. These resistor netlist entries also provide a 0.5% correlated variation. Therefore, each resistor will have values that vary within ±10% of their respective nominal values, but they will vary (relatively, in the ratio 47:10) from each other within ±0.5%.

The random function

The MAST language provides an intrinsic function, **random()** that returns a pseudo-random number in the interval that includes 0 and goes up to, but does not include, 1. This function uses no arguments.

The **random()** function is useful for simulating something with a certain probability. For example, assume you want to flip a coin (i.e., have a variable that takes on two discrete values with certain probabilities). Although this could be described with a piecewise linear distribution, it is simpler to use the **random()** function. The following example shows how to define a parameter that has 0.4 probability of being 1k, and 0.6 probability of being 2k:

```
number r, p
p = random()
if (p<.4)    r = 1k
else         r = 2k
```

12.4 Noise Modeling

The Saber simulator can perform a noise analysis to include the noise contributions of a circuit or system element. To do this, the template must contain information that defines its noise contribution.

In general, a noise source for an electrical element is defined either as a current or voltage source between two nodes of the element. A simple element, such as a resistor, uses a single noise source. For a more complex element, such as a transistor, there may be several noise sources, as well as several types of noise: thermal noise, shot noise (due to DC current), and flicker noise (a frequency-related noise).

Adding noise information to a template is not difficult, and the procedure is the same for all types of small-signal noise. In general, it consists of the following:

- Define the name of the noise source as a **val** (a local variable).
- Provide the defining expression for the noise **val**.
- In the control section, insert a **noise_source** statement that supplies one of the following kinds of information:
  - If the noise source is a current source, the statement describes the location of the noise source in terms of the connection points or internal nodes.
  - If the noise source is a voltage source, the statement associates the name of the noise source with a **var** (an independent system variable).

If there is more than one noise source, the control section must contain a separate variable, definition, and statement for each. In this chapter, noise information is added to the simple **resistor** template of Chapter 7 and the simple **diode** template of Chapter 9.

Adding Noise to a Resistor

For this example, only the thermal noise through the resistor is defined. This noise source is defined as a current source in parallel with the resistor, as shown in Figure 12-8. Note that there is no direction associated with the current source.

To include thermal noise effects in the template, the following expression is used to define them:

$$noise = \sqrt{\left|\frac{4kT}{r}\right|} \qquad (12\text{-}11)$$

where

k is Boltzmann's constant (1.38×10^{-23} joules/Kelvin)

T is the temperature in Kelvins

r is the specified resistance

Advanced AHDL Capabilities

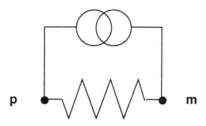

Figure 12-8. Defining a noise generator.

Note that it is necessary to declare variables for Boltzmann's constant and temperature, in addition to the noise source variable.

The variable for simulation temperature (**temp**, in °C) is an externally defined number. The unit for thermal noise (**ni**) generated by a current source is predefined as A/√Hz. By declaring a noise variable as a **val**, the unit **ni** (A/√Hz) is assigned to it. For example, a noise variable named **nsr** would be declared as follows:

 val ni nsr

The constants used to calculate noise, listed above for Eq. (12-11), must also have local declarations:

 number k = 1.38e-23
 number t

Because the value defined externally for **temp** is in °C, a statement is required to express the temperature (**t**) in Kelvins:

 t = temp + 273.15

Using these variable names for the noise generator, insert a statement to perform the noise calculation shown in Eq. (12-11):

 nsr = sqrt(abs(4.0*k*t/res))

The last thing required is to insert a **noise_source** statement in the control section. It identifies the noise source in relation to the rest of the template. If the noise source is a current source, as in this template, the statement contains the name of the pins (or internal nodes) to which the noise generator is connected. If one side of it is connected to ground, only the other need be listed, in which case the simulator assumes that the other side is grounded.

In this example, as shown in Figure 12-8, the noise source **nsr** is connected between pins **p** and **m**. Therefore, the statement is the following:

> **noise_source (nsr, p, m)**

This statement adds the noise to **p** and subtracts it from **m**. Alternately, because the noise analysis ignores the sign of the noise source, the following statement would be an equivalent statement:

> **noise_source (nsr, m, p)**

For a noise current source, the general form of the **noise_source** statement in the control section is as follows:

> **noise_source** (*val_name, pin [, pin]*)

For a noise voltage source, the form of the statement would be as shown below, where *var_name* is the name of a **var** defining the current through the voltage source:

> **noise_source** (*val_name, var_name*)

With all noise-related statements added, the **resistor** template becomes the **resistor_ns** template shown below.

Adding noise to the diode template

The **diode** template of Chapter 9 defines a simplified diode model. Shot noise is incorporated for illustration.

The shot noise is defined as a current source, and it is connected between pins **p** and **n**. The defining equation for shot noise is as follows:

$$nsi = \sqrt{2q \cdot |id|} \qquad (12\text{-}12)$$

where q is the electron charge and id is the current through the diode.

Advanced AHDL Capabilities

```
template resistor_ns p m = res
electrical p, m
number res
external number temp         # external parameter added for noise
{
branch cur=i(p->m)
branch vlt=v(p,m)
val ni nsr                   # val added for noise
number k=1.38e-23            # local parameter added for noise
number t                     # local parameter added for noise
t=temp + 273.15              # calculation added for noise
nsr=sqrt(abs(4.0*k*t/res))   # calculation added for noise
control_section {            #control section added for noise
    noise_source (nsr, p, m)
    }
cur=vlt/res
}
```

Therefore, only three statements need to be added to this template to define shot noise:

- Define the name of the noise source as a **val**:

    val ni nsi

- Provide the defining expression for the noise variable:

    nsi = sqrt(2*qe*abs(id))

- In the Control section, insert a **noise_source** statement that associates the name of the noise source with a **var**:

    noise_source(nsi, p, n)

From this example, it should be clear that adding noise information is a very straightforward process, regardless of the complexity of the template. Adding the other noise information (thermal noise and/or flicker noise) is simply a matter of defining each necessary variable, adding its defining expression, and inserting its noise_source statement in the control section of the template.

Conclusions

The advanced features that have been described in this chapter are not necessarily the most often used aspects of an AHDL, but they become extremely important as the designer attempts to use computer simulation to investigate phenomenon impractical or impossible to observe by any other means (such as breadboarding). Other areas such as worst case analysis, failure effects, and design for reliability are studies into which simulation offers a tremendous amount of insight.

An further example of this enhanced simulation capability is the modeling and simulation of transient noise. Using the modeling concepts introduced in this and the previous two chapters, transient (large-signal) noise effects can be modeled. Specifically, the statistical functions and the event-driven constructs (when statements, states, etc.) can be used to create the signature of the noise. This noise is then simply added to the appropriate signals. This is very useful when simulating mixed analog-digital systems where a small-signal noise analysis cannot be performed.

12.5 References

1. L. B. Jackson, *Digital Filters and Signal Processing*, 2nd Edition, Kluwer Academic Publishers, 1989.
2. R. Dorf, *Modern Control Systems*, 3rd Edition, Addison-Wesley, 1980.
3. A. V. Oppenheim and A. S. Willsky, *Signals and Systems*, Prentice-Hall, 1983.

Section 3

Advanced Applications

Chapters 13-18

This section contains several examples of designs that use MAST models from non-electrical technologies. Each example makes use of concepts brought out in the first two sections. The main purpose of these chapters is to illustrate the importance of using an AHDL to enhance the power of computer simulation.

Each example is accompanied by a "pseudo-template" for a selected model that is representative of the technology. The purpose of this is to outline the necessary details for implementing a given model. This type of template is not executable as shown; it only serves to compare the basics of electrical and non-electrical models.

CHAPTER 13

Electro-mechanical Systems

Most of the modeling concepts described thus far have been focused on their application to electrical circuit simulation. This is only natural, since this discipline has driven the development of analog hardware description languages. There are, however, other applications and technologies that can be naturally encompassed by an AHDL. The remaining chapters illustrate many of the possibilities to which modeling languages can be effectively applied. There are many more such areas than are presented here, so these chapters are only a representation.

This chapter introduces models for electro-mechanical systems, or *mechatronics*. In order to model and simulate mechatronic designs, it is necessary to not only represent electrical components, but mechanical ones as well. However, mechanical components can be described in similar ways to electrical. Fundamentally, the same mathematics are employed—ordinary differential equations. In simulations of strictly mechanical systems it is common that the mechanical engineer is interested in internal forces and effects of the mechanical components. This has led to the use of finite element (FE) simulation of mechanical parts. However, the same is true of electrical devices. Once either the mechanical or the electrical engineer proceeds to the systems level however, this is no longer the level of detail appropriate for simulation. Thus, for simulation of mechatronics, this book will focus on one-dimensional models (as opposed to two- and three-dimensional models found in FE models) for the electrical and mechanical components that comprise the system. The techniques described in the first two sections of this book for electrical models apply equally to mechanical models.

The example in this chapter is that of an automotive seat position controller [1]. It includes electrical, mechanical and thermal effects.

13.1 Overview of Mechanical Modeling

There are three basic types of mechanical templates: translational (linear or straight-line motion), rotational (circular motion), and combinations of translational, rotational, or other types (e.g., electrical, switch state, etc.).

The variables and connection points of mechanical templates are defined much like they are in electrical templates. This has the following implications for simulating mechanical systems:

- Each mechanical template has pin-type connection points.
- Mechanical systems, like electrical circuits, require a reference node (0).
- Conversion templates must be used when connecting templates that have different through or across variables, such as mixing electrical and mechanical.
- The simulator will solve for through and across variables at nodes connecting templates. For instance, Table 13-1 compares through and across variables for electrical and mechanical templates.

Simulators such as Saber and SPICE solve for the steady-state or transient solution of a network by finding the instantaneous values of the across variable such that the sum of the through variable, at each node or connection point, equals zero. This is known as Kirchhoff's Current Law (KCL) in the electrical domain, but the concept applies equally to the mechanical

Electro-mechanical Systems **175**

domain. That is, the sum of the forces (or torques) on a body equals zero.

Newton's law (F = m•a), then, can be restated in the same form as KCL: if the force due to the inertial acceleration (m•a) is accounted for on the left side of the equation, then F_{appl} - m•a = 0.

Table 13-1. Comparison of electrical and mechanical connections.

	Mechanical		Electrical
	Rotational	**Translational**	
through variable	torque	force	current
across variable	angle or angular velocity	position	voltage

Most of the mechanical models contribute force or torque at their connection points. The magnitude of the through variable (force or torque) is generally a function of the across variable (position or angle) seen at the connections (or their derivatives). The form of this function is determined by the physical phenomenon being modeled. For example, a translational spring contributes a force that is proportional to the relative position between its two connections. Similarly, a viscous damper contributes a force that is proportional to the relative velocity (derivative of the position).

The through variable polarity chosen for these models is such that a positive force or torque applied at a node will produce a positive increase in the across variable (position or angle) at that node. For example, if a torque source is connected to a moment of inertia, a damper and a spring at node **a1**, then a positive torque value specified for the source will cause the angle seen at **a1** to increase. In response to that increase, however, the mass, damper and spring will all contribute negative torques (proportional to the angular acceleration, angular velocity and angle, respectively).

The across variable represents an instantaneous value or quantity that is common to all templates connected to that node.

For example, in translational systems, the "position" across variable at a node represents the physical location of all templates connected there. For most models, it is the position difference between their two connection points that determines its effect (i.e., the through variable contribution) on the system.

When analyzing a rotational system, it is possible to use either the angle of the rotating body, or the rate at which it is turning (its angular velocity), as the across variable. Consequently, in some cases two distinct templates are needed to model the same physical phenomenon:

- one with angle as the across variable
- one with angular velocity (ω) as the across variable

It is important to remember that using templates with different across variables in the same system requires a conversion to be performed—usually by an additional template inserted at the interface. The template **conv_r** is provided for this purpose. It allows using models with different types of rotational across variable (angle or angular velocity) within the same design.

System of units

Various systems of units for describing mechanical dynamics are in common use. Predefined systems of units are available to accommodate both metric and English units, while providing large- or small-scale dimensions within each system. It is possible to create a template that has more than one system of units available [2].

Examples of such units are:

metric	MKS (meter, kilogram, second)
	CGS (centimeter, gram, second)
English	FPS (foot, pound, second)
	IOS (inch, ounce, second)

Implementing mechanical models

Implementing mechanical models with an AHDL is quite similar to that of implementing electrical models. There are two main differences. The first is that there are alternative through and across variables as described in Table 13-1 in the last section. The function of the mechanical device determines the required connection type. The second difference in implementing mechanical models is the various systems of units in use. This choice is typically left to the modeler. The most common system is MKS. Conversions are typically performed to other systems from this base system of units.

In the *pseudo*-template shown below of a rack-and-pinion device, the necessary implementation details for mechanical models is outlined. This template is not executable as shown. It only serves to highlight the key differences between mechanical modeling and electrical modeling. Of course, creating an electro-mechanical model is covered by this description as well since nothing prevents the use of both technologies within one template.

The rack-and-pinion mode **rack_pin** illustrates the use of both rotational and translational mechanical connections. In this particular *pseudo*-template, the rotational connection declaration indicates that the angle will be used as the across variable as opposed to angular velocity. If angular velocity was the desired across variable, the declaration would read as follows:

rotational_vel ang

Of course, the through variable for any rotational connection is torque. The translational connection type in this case uses position as the across variable and force as the through variable.

The units that the arguments are assumed to have upon designation must be selected through reference to a predefined set of units (which can be appended with additional units). These units are then used in the declaration of local variables as illustrated in the local declarations portion of the *pseudo*-template.

The remainder of the implementation process for mechanical models is completely analogous to that of electrical modeling. The parameters are bulletproofed for validity, one-time calculations may be performed on them, and the governing equations are implemented (either through the use of the values and equations sections or in simplified MAST). Of course, there are different operators for mechanical connections than for electrical. The angular position **ang_rad** and translational position **pos_m** operators are shown in the values section. The angular velocity operator is **w_radps**. These across operators are analogous to the v() notation for electrical templates. The torque and force operators are shown in the equations section as **tq_Nm** and **frc_N**, respectively.

```
template rack_pin ang pos = arg1, arg2, units
rotational_ang ang           # Mechanical angular connection (pinion).
translational_pos pos        # Mechanical position connection (rack).
number arg1, arg2            # Declare arguments of template as before
mech_def..units units = mks  # Selects units assumed for arguments.
{
# Local declarations as performed for electrical templates except for units
val ang_rad   angle_mks      # Pinion angle (radians).
val tq_Nm     torq_mks       # Pinion torque, Newton*meters.
val pos_m     posn_mks       # Rack position (meters).
var frc_N     force_mks      # Rack force, Newtons.
parameters{
    # Perform one-time parameter processing as in electrical templates
    }
values {                     # Implement governing equations
    angle_mks = ang_rad(ang) # Mechanical operator ang_rad analogous to v( )
    posn_mks = pos_m(pos)    # Mechanical operator pos_m analogous to v( )
    torq_mks = -1 * radius * force_mks
    }
equations {
    tq_Nm(ang) += torq_mks
    frc_N(pos) += force_mks
    force_mks: radius * angle_mks = posn_mks
    }
}
```

13.2 Automotive Seat Position Controller

In the past, electrical engineers would design and validate their circuits using a general electrical simulation tool, or some other software tool that was used only for a specific part of the design. Mechanical, thermal, and digital engineers would also use their specific software tools for validation of their designs. It was not easy to link these designs into one simulation and validate compatibility of the entire system. This example illustrates a practical engineering design problem incorporating thermal protection in an electro-mechanical motor drive system.

The seat position motor drive circuit is shown in Figure 13-1. This circuit contains six models which provide an integrated thermal, electrical, and mechanical simulation validation of the total design. The automobile battery powers the system when the switch is closed. The switch represents the button on the side of the car seat that activates the seat position motor. Once activated, the motor drives the gears, which in turn drive a rack-and-pinion system within mechanical limits. The temperature-limited MOSFET is in series with the armature winding of the motor to provide thermal protection. This thermal protection circuit uses an integrated electrical/thermal transient model of a power MOSFET with temperature sense and shutdown circuitry.

Mechanical models

The motor is modeled as a permanent magnet DC machine that may act as either a motor or generator depending on how it is connected. Machine parameters include the resistance and inductance of the armature, interpole and compensating windings. The torque and back emf constant are numerically equal to each other and are modeled with a single parameter. Internal damping and inertia of the unloaded machine, as well as the number of machine poles, are additional parameters. Motor models will be described in further detail in the next chapter.

The gear model represents a simple mechanical gear. Its behavior is such that a fixed relationship is maintained between the shaft 1 and shaft 2 angular velocities.

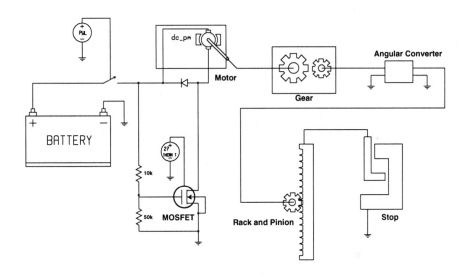

Figure 13-1. Seat position motor drive with active thermal cut-out.

That is, the velocity of shaft 2 is equal to the velocity of shaft 1, scaled by the gear ratio gr as in Eq. (13-1):

$$velocity(shft2) = velocity(shft1) \cdot gr \qquad (13\text{-}1)$$

This velocity relationship is maintained by the model by generating the necessary torques on the two shafts. Power is conserved by the gear model, as the shaft torques are also scaled by the gear ratio:

$$torque(shft1) = -torque(shft2) \cdot gr \qquad (13\text{-}2)$$

Note the minus sign in the torque equation. This is because the reaction torque applied back on the *drive* shaft is opposite to the torque applied by the *driven* shaft.

A conversion model is used as an interface between the gear model and the rack-and-pinion model. It takes a positive angular velocity from the gear model and converts it into an increasing angle value of the pinion for the rack-and-pinion model.

The rack-and-pinion model acts as a rotational/translational converter, with mechanical rotation angle and translational position connections. Its behavior is such that a fixed relationship is maintained between the pinion rotation angle (**ang**) and the rack translational position (**pos**) as:

$$radius \cdot sign \cdot angle\,(ang) = position\,(pos) - pos0 \qquad (13\text{-}3)$$

Note that the scale factor radius (which represents the radius of the pinion), and a zero angle position offset pos0 are provided. Note also that a sign factor, equal to +1 or -1, allows positive rotation angle changes to cause an increase or a decrease in the position, respectively. This defined position/angle relationship is maintained by the model by generating the necessary forces and torques (on the rack and the pinion, respectively) to enforce the relationship. The products of torque times angular displacement, and force times linear displacement, are balanced by the rack-and-pinion model (so that energy is conserved). This occurs because the torque and force are also scaled by the factor radius:

$$torque\,(ang) = -radius \cdot sign \cdot force\,(pos) \qquad (13\text{-}4)$$

Again, note the minus sign in this equation. This is because the reaction force (torque) that is applied back on the *driving* member is opposite to the torque (force) applied by the *driven* member.

The mechanical stop model represents a two-position hardstop (mechanical limit). Its behavior is such that when the relative position (**pos1** - **pos2**) is between the limits **ptop** and **pbot**, the model exerts no force. However, when the position exceeds either limit (that is, if the relative position increases above **ptop** or drops below **pbot**), then a force proportional to this excess is generated. The force polarity is such that the relative position is driven back into the non-limiting region. The proportionality constants (i.e., the mechanical "stiffness") for the top and bottom hardstop limits are given by the parameters **ktop** and **kbot**, respectively. In addition, an energy dissipation factor is available, to model the inelastic effects of collisions with the hardstop. The damping factors, **dtop** and **dbot**, provide this energy loss effect.

Modeling the temperature-limited MOSFET

The MOSFET of Figure 13-1 uses the temperature effects on internal threshold voltage (V_T) and forward diode voltage drop (V_d) to provide the triggering mechanism versus temperature. Positive feedback as well as a hysteresis loop is also included in the model. Figure 13-2 shows these characteristics of the MOSFET model. This figure shows the junction temperature of the device being forced to increase from 25°C to 200°C. The results show the device quickly shuts off at approximately 175°C and then turns back on at approximately 135°C. This validates the hysteresis loop, the positive feedback, and the temperature triggering circuit of the MOSFET model. More details of modeling self-heating effects will be given in Chapter 16.

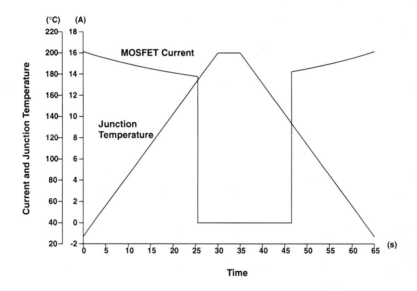

Figure 13-2. MOSFET current and junction temperature illustrating thermal hysteresis.

Integrated seat position system simulation

Figure 13-3 shows the system level results of all the thermal, electrical, and mechanical models working together. The simulation objective is to activate the seat position motor and move the position of the seat until the seat hits the physical stops.

Electro-mechanical Systems **183**

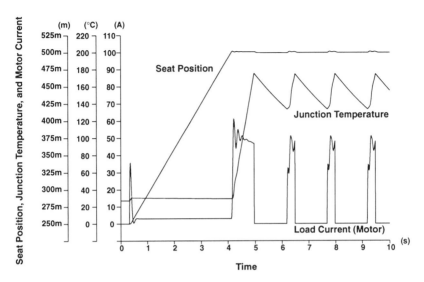

Figure 13-3. Seat position, junction temperature and motor current of seat position system.

Normally, the person holding down the button for the seat to move would release the button at this time, and the motor would shut off. However, the simulation assumes that a fault occurs (as when a child is playing with the seat position and does not release the button when the seat hits the physical stops). With the seat up against the stops and the motor button still engaged, the motor is forced into a stall condition. At this point the motor's armature current increases. The temperature limited MOSFET device model is in series with the armature winding, and therefore carries this increased current. This causes the MOSFET to heat up until the junction temperature rises to 175°C. At this temperature the MOSFET circuit shuts the device down and thus the motor shuts off. With no current flow through the MOSFET, it begins to cool down.

If the seat position button is still activated when the MOSFET device model cools down to 135°C, the MOSFET will turn back on and the cycle will repeat.

Conclusions

With this example it is possible to see how an AHDL allows one to mix systems of differing types and focus on the interactions of these technologies. In the above example it is observed that adding insulation to the temperature-limited MOSFET would extend the temperature decay beyond the one second range, so that the device would remain off longer—a conclusion not necessarily obvious at the outset.

13.3 References

1. S. Chwirka, J. M. Donnelly, and H. A. Mantooth, "Integrated thermal, electrical and mechanical simulation of an automotive seat position system," *PCIM Magazine*, pp. 26-29, Nov. 1993.

2. *Mechanical/Motor Template Libraries*, Pub. MB-0147, Analogy, Inc., April 1993.

CHAPTER 14

Motors

The simulation of complex designs, whether analog or mixed analog-digital, is sufficiently problematic considering convergence issues, analysis capability, time constraints, etc. without considering the lack of design coverage in available models. This is true without even leaving the electrical domain with most simulation tools. SPICE [1] and SPICE-based simulators have the added problem that it is very difficult to add models to the simulator. Indeed, modeling quickly becomes the biggest impediment for people really trying to utilize simulation tools. Once outside of the electrical domain, AHDLs make a vital impact on modeling. While macromodeling has a lot of merit in the electrical domain, macromodeling non-electrical devices with electrical equivalents is more difficult and quite confusing.

One of the areas where AHDLs make a major impact is the modeling of electrical machines (i.e., motors and generators). This chapter is focused on the subject of modeling motors. The first section provides a short overview of modeling motors. The next section is devoted to implementation details of motor models in MAST. The remaining sections are devoted to the description of an example illustrating the use of a motor in a system simulation.

14.1 Overview of Motor Modeling

A motor or generator is by its very nature or function an electro-mechanical device. Thus, the connection points of motor models will typically consist of electrical inputs and rotational mechanical outputs. For the mechanical connections, the through and across variables are torque and angular position, respectively.

All motors provide a conversion between electrical and mechanical energy. That is, electrical voltage and current at the input terminals cause rotational speed and torque on the output shaft (or vice versa for a generator). Further, this conversion is based on the principle (Faraday's Law) that a loop of conductor carrying current within a magnetic field will result in a mechanical force exerted on the conductor, if the magnetic flux passing through the loop is changed. Properly coupling this conductor to a shaft provides the output torque.

Thus, there are two necessary elements for any motor: (1) the magnetic field—provided either by a permanent magnet (PM) or by an electromagnet (wound-field), and (2) the current-carrying conductor (which is referred to as the armature winding). Often, the element that remains stationary is called the stator, and the one that is rotated is called the rotor.

There are several basic types of motors, including DC motors, induction motors, and synchronous motors. An overview of all of the technical details of the various types of motors is beyond the scope of this book. References are available on this subject [2, 3]. However, each type of motor has been implemented in MAST and the example in the next section will demonstrate the use of one such model.

Implementing Motor Models

Implementing motor models in MAST is simply a combination of the constructs learned to implement electrical and mechanical models. The motor is simply an electro-mechanical device and as such will possess electrical inputs and mechanical outputs. The *pseudo*-template shown below is an outline of the aspects of a motor model that must be addressed. This template is an outline of the stepper motor model used in the example application in the next section.

Motor templates consist of electrical input connections and rotational mechanical output connections. The *pseudo*-template for the stepper motor **stp_2ppm** illustrates this.

```
template stp_2ppm ap am bp bm thetarm =arg1, arg2, units
electrical ap, am, bp, bm           # Electrical connections
rotational_ang thetarm              # Rotational mechanical connection
number   arg1=0.1,
         arg2=undef
mech_def..units units=mks           # Reference desired mechanical units
{
# Declare internal vals (some electrical and some mechanical)
parameters{
    # Perform one-time parameter processing...
    }
values {
    # Implement governing equations...
    }
equations {
    # Implement topology of model such as:
    i(ap->am) += ia                 # Branch currents
    i(bp->bm) += ib
    tq_Nm(thetarm) += tq_mks        # Output torque
    }
}
```

14.2 Floppy Disk Drive Head Position Controller

The application that is presented to illustrate the use of motor models is a floppy disk drive system. A block diagram of the floppy disk drive system is shown in Figure 14-1 [4]. The system contains a microprocessor, which activates the spindle motor controller, the stepper motor controller and the read/write control. The spindle motor controller and the stepper motor controller are ASICs that are specifically designed for motor control. The read/write control accesses the data on the floppy disk for reading or writing.

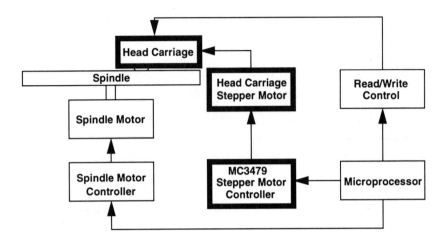

Figure 14-1. Block diagram of a typical floppy disk drive system.

A portion of this system, the drive head position control, was selected to demonstrate the use of motor models in a mixed technology simulation. This portion of the system is indicated with the bold boxes in Figure 14-1, and includes the MC3479 stepper motor controller chip [5], the head carriage stepper motor, and the head carriage assembly.

The drive head position control schematic is given in Figure 14-2. On the left side of the schematic the digital command signals from the microprocessor enter the MC3479 stepper motor driver chip, a mixed analog-digital ASIC. The function of the chip is to act as an interface between the microprocessor and the mechanical system that controls the read/write head position of the floppy disk. The signals incident from the microprocessor are the output impedance control (OIC), direction control (CWB), step control (FULLB) and the clock signal (CLK). Immediately beneath the driver chip is a three-resistor and one transistor power-saver circuit that generates the proper input bias current (BIAS) for the MC3479. The maximum drive current of the MC3479 is proportional to this current. When the power-saver circuit is activated by a high signal on the base of the bipolar transistor, the maximum output drive current is generated. When deactivated, this current is one-half the maximum, thus saving power.

Motors

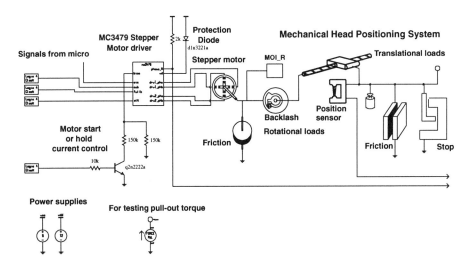

Figure 14-2. Schematic diagram of the floppy disk drive head position controller.

From the perspective of the ASIC, the remainder of the schematic represents the load that the chip must drive. The primary outputs of the driver chip are connected to a two-phase stepper motor. The stepper motor drives a lead screw head positioner. Many of the other elements shown on the schematic in the mechanical section represent nonideal aspects of the motor and the lead screw head positioner. The three elements between the motor and the lead screw are rotational loads on the motor in the form of friction, backlash from the lead screw, and rotational moment of inertia. Among the four elements shown after the lead screw that are attached to the disk drive head are the position sensor and translational loads on the lead screw. These translational loads are the mass of the head, friction between the screw and the head, and a mechanical stop.

Modeling of stepper motor driver ASIC

The stepper motor driver chip was represented hierarchically in this application. It was broken down into its constituent parts and each part modeled separately. Some of these blocks were purely digital, some were purely analog and others were mixed analog-digital models. Most of the digital models were performe at the gate level. However, a state machine was also modeled at Boolean logic level [4].

The output drive stage of the chip was represented by th transistor-level circuitry actually employed in the design. Thi block of circuitry could have been replaced by a behavioral mode that would allow the simulations to be more computer efficient However, since the interface between the motor and the drive chip was of interest, this was the most appropriate representa tion.

Modeling of stepper motor

A stepper motor rotates only a specific fraction of a revolutio on command. This enables it to control the position of its load ver accurately. There are three basic kinds of stepper motors: reluc tance, permanent magnet (PM), and PM-hybrid. The steppe motor in this example is a PM stepper motor. The input to a step per motor determines shaft position; thus, the model uses angula position (as opposed to velocity) as its across variable. The basi configuration of a stepper motor is shown in Figure 14-3.

Stepping action occurs as follows. Assume that the b windin (bp-bm) is open-circuited and a constant positive current is flow ing in the a winding (ap-am). The rotor would be positioned at q 0. Simultaneously de-energizing the a winding while energizing with positive current causes the rotor to move one step length i the clockwise direction. To continue stepping in this direction, b i de-energized and a is energized with a negative current. Thus, i is necessary for currents to flow in both directions to achieve rota tion.

The stepper motor model of Figure 14-3 is a behavioral mode of a two-phase, permanent magnet stepper motor with four elec trical inputs and a mechanical output that is the angular positio of the shaft. This model assumes constant self-inductances and constant reluctance seen by the permanent magnet, independen of rotor position. This disregards the reluctance torques (cause by variations in self-inductances) that attempt to place the roto in its minimum-reluctance position. However, the model does pro vide a parameter (**kd**) that allows the user to specify a reluctanc torque caused by the magnet—the detent (or retention) torque

Motors

This torque exists regardless of whether the stator windings are

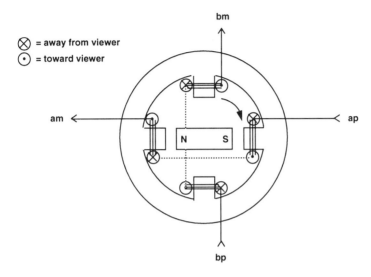

Figure 14-3. Two-phase PM stepper motor.

energized, and it will preserve the rotor position during a power failure (if the load torque is not too large). Other parameters of the model include winding resistance and inductance of the two phases. Torque constants for both electromagnetic torque, as well as detent torque (torque required to move the shaft from a stationary state), are also parameters. Internal damping and inertia of the unloaded machine, as well as the number of machine poles, are additional parameters.

The following is a typical netlist usage of a characterized stepper motor model [6]:

stp_2ppm.1 w1 w2 w3 w4 thetarm = r=75, l=60m, km=100m, kd=250u, d=250u, j=1.7u, p=100

This motor model provides a more realistic model of the load of the MC3479, since it models the back emf of the motor as a function of the load on the motor. Thus, the other mechanical elements do have an influence on the electrical performance of the ASIC through the motor model. The inductance of the motor is also variable.

Modeling the mechanical head positioning system

The rotational friction model connected to the shaft of the motor has two angular connection points. This model includes the effects of stiction, or static friction, that is a common phenomenon in mechanical devices. Basically, the model will exert a torque between its connection points that is a function of the relative angular velocity between the connections. If the relative velocity is above **velmin** (a model parameter), the model exerts a constant torque equal to the friction value **fric** (another model parameter). If the velocity is below **velmin**, the torque approaches the value **stic**, the model parameter for stiction. When the velocity becomes 0, the model exerts a torque as needed to hold the present angle, but this torque is limited to the value of **stic**.

The rotational moment of inertia model is a mechanical primitive that models a reaction torque proportional to the relative angular acceleration of the shaft. The only model parameter is the proportionality constant.

The backlash symbol of Figure 14-2 represents a model of the rotational backlash (i.e., "slack" or "play") of the lead screw, which is common in many mechanical devices. Its behavior is such that the relative angle between the connections (ang1 - ang2) is maintained between **-dtheta/2** and **dtheta/2**, where **dtheta** is a model parameter. If the relative angle exceeds these limits, a torque proportional to the excess torque is applied, directed so as to bring the angle back within the limits. While the relative angle is in the backlash region, no torque is transmitted between the connections, except perhaps for a preload (or "antibacklash") spring torque, if one is specified.

Next is the model of the lead screw head positioner. The input connection of this model is a rotational angle and the output is a translational position. The fundamental equation describing the torque-force relationship is:

$$F = 2\pi \cdot T \cdot p \tag{14-1}$$

where T is the applied torque and p is the pitch of the lead screw. Most of the non-idealities of the lead screw model are implemented as external components as shown in Figure 14-2.

Motors 193

The position sensor is another behavioral model used to detect the position of the head. The other symbols connected to the head are translational loads. These include the mass of the head, the friction of the head against the lead screw, and the mechanical stop. The friction model is very similar to that described for rotational friction, with the main differences being the conversion to translational.

The mechanical stop exerts no force when the relative position is between the stops. When the stop is reached, a force proportional to the excess force is generated. The force polarity is such that the relative position is driven back into the non-limiting region. In addition, an energy dissipation factor is available, to model the inelastic effects of collisions with the hardstop. Damping factors provided as model parameters provide this energy loss effect.

Simulation results

The first simulation result is that of the settling time of the system in order to determine the access time of the floppy drive. The signal shown in Figure 14-4a is the angular position of the motor shaft as a function of time. The excitation to the motor was that necessary to force it to advance a single step. The settling time is on the order of 45 ms. Figure 14-4b shows the same shaft angular position, but in this case the motor was running at full speed and brought to an abrupt stop. In this case, the motor shaft settles after about 85 ms. This is a limiting factor for the access time of the overall floppy drive system. During this simulation the electrical portion of the system is operating under maximum drive current conditions, which is well within normal operation.

Another interesting result is a phenomenon known as "loss-of-step" for the motor. This condition occurs when the motor is unable to respond quickly enough to the electrical stimuli applied to its inputs before the next stimulus arrives. One such cause might be that the motor is overloaded. Figures 14-5 and 14-6 illustrate a couple of examples of loss-of-step. In Figure 14-5a an impulse of force is introduced into the system at 62 ms. Depending on the magnitude of this impulse, the motor will experience loss-of-step. In this case all impulsive forces less than 10 N do not cause this condition. The top two curves represent impulsive

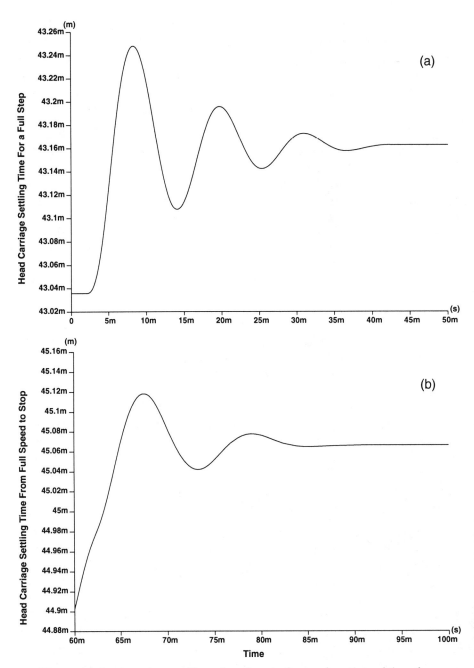

Figure 14-4. Angular position of motor shaft as a function of time for (a) single step advance (b) full-speed to stop.

forces of 15 N and 20 N, respectively. In each of these cases the motor has lost step. In Figure 14-5b the back emf generated in the motor when it loses step is shown.

In Figure 14-6 the angular position of the motor shaft is again displayed versus time. In this sequence of curves, the variable is the mass of the head carriage while the motor is started from stop, accelerated to full speed (~35 ms), reversed, reversed again (~65 ms), reversed a third time (~100 ms), and brought to full stop (~130 ms). As the mass is increased up to 20 g, it is observed that the critical mass at which the motor loses step is between 15 g and 20 g. The 20 g simulation loses step when the motor is reversed for the second time at about 65 ms. This mechanical condition does not pose any severe operating conditions on the electrical system. However, it does confirm that if the electrical drive of the motor is too fast for the given mechanical conditions, the simulations will predict when loss of step will occur.

Conclusions

The first question that must be answered regarding simulation is: What *must* the designer be able to simulate in order to insure that the design will work as specified? As the system complexity grows, it becomes necessary to employ alternative methods of representing a design in order for simulation to yield meaningful results in a timely fashion. The mechanical load in the system of Figure 14-2 could be represented by some simple electrical load and in some instances this is quite acceptable. However, as the design limits are pushed or additional performance is required, a more realistic model is needed. The designer faces a dilemma without an AHDL—the inability to make an easy transition from a lumped electrical load to a more realistic dynamic load. Many generic models (e.g., motors, sensors, friction, etc.) have already been developed in MAST. The modeling language allows the designer to create models as required. This allows the designer to move to the next question: What would I *like* to simulate to further improve my design?

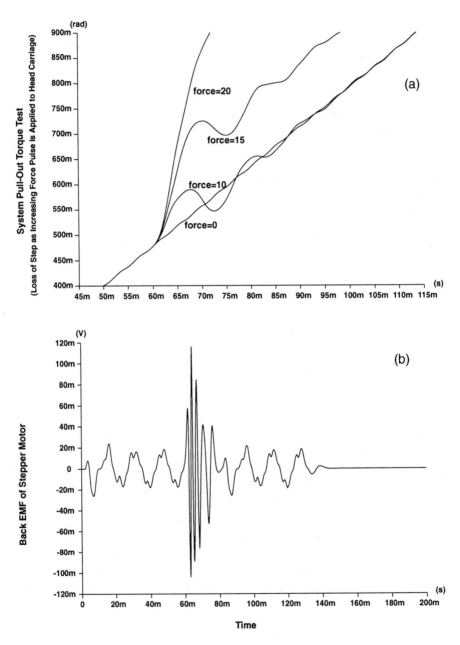

Figure 14-5. Loss-of-step of stepper motor (a) angular position of shaft for impulsive forces of increasing magnitudes (b) back EMF.

One of the biggest factors in reducing time-to-market in any design is the utilization of concurrent engineering whenever possible. The ability to represent each aspect of the system in its native units and with a variety of complexity makes it possible for the mechanical engineer and the electrical engineer to design this system concurrently and share models as part of the design process. Of course, the same holds for analog and digital designers within the electrical domain.

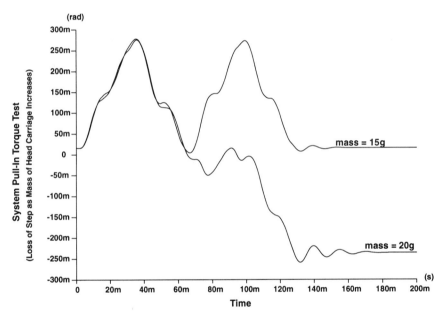

Figure 14-6. Angular position of motor shaft illustrating loss-of-step as a function of the mass of the head carriage.

14.3 References

1. L. W. Nagel, "SPICE2: A computer program to simulate semiconductor circuits," *ERL Memo No. ERL-M520*, University of California, Berkeley, 1975.
2. G. R. Slemon and A. Straughen, *Electric Machines*, AddisonWesley, 1980.
3. P. C. Krause, O. Wasynczuk, and S. D. Sudhoff, *Analysis of Electric Machinery*, McGraw-Hill, 1986.
4. J. R. Carlson and H. A. Mantooth, "Simulation of a floppy disk drive head position controller," *Analog Circuit Design - Mixed A/D Circuit Design, Sensor Interface Circuits and Communication Circuits*, pp. 53-67, Kluwer Academic Publishers, 1994.
5. MC3479 Stepper Motor Driver, Motorola Semiconductor Technical Data Book, 1990.
6. SST39D1041 Stepper Motor, Shinano Kenshi Motor Data book, Catalog No. SST-93, 1993.

CHAPTER 15

Hydraulics

15.1 Overview

Hydraulic elements and systems have many similarities in behavior to those of electrical devices and circuits. Because hydraulic behavior can be expressed as ordinary differential equations for continuous-time simulation, the same types of analyses may be performed on hydraulic models as those that are done on electrical models. Consequently, the MAST AHDL allows straightforward implementation of hydraulic templates, using the same constructs and procedures for electrical templates introduced in Chapters 1 through 12.

As with models for other non-electrical technologies (such as mechanical and magnetics), once the through and across variables have been determined, the unit and pin definitions for hydraulic templates can be implemented accordingly. This allows full description of governing equations for the technology.

Hydraulic models are based on conservation of flow. That is, the simulator solves for the node pressures such that the total flow into any given node sums to zero. Each device connected at that node can contribute flow, based on an absolute or a differential pressure. Models that have the property of volume account for flow due to fluid compressibility. Other dynamic effects, such as fluid momentum, are accounted for in the models with a moving fluid column. In addition, nonlinear effects, such as laminar vs. turbulent flow and fluid cavitation, may also be included.

15.2 Fundamental Hydraulic Quantities

Note that hydraulic "compressibility" differs from electronic "capacitance" in an important aspect. The charge on a capacitor is a function of the *relative* voltage across the capacitor. In contrast, the compression of a hydraulic fluid is a function of the *absolute* pressure at each pin. For example, the charge across a capacitor would be zero if the voltage at each pin of a capacitor were 1 (or 1 million) volts, as long as there was no difference in the voltage at the pins. On the other hand, the fluid in a hydraulic tube would compress if the pressure at each pin were to increase.

Basic hydraulic devices to be modeled include:

- Sources (flow, pressure)
- Storage devices (accumulators)
- Losses (orifices, porous membranes, lines)
- Transducers (actuators, motors, pumps, radiators)
- Controllers (switches)
- Model interfaces to other disciplines (such as control systems)

Flexible transmission line model

Figure 15-1 shows the symbol for a flexible hydraulic transmission line, followed by a *pseudo*-template (**hose**) that demonstrates portions of this model. Note that the declarations for pins **p1** and **p2** (**hyd_mks**) includes the MKS (meter-kilogram-second) system of hydraulic units. The across variable is pressure (**p_Npm2**) in newtons per square meter. The through variable is flow (**q_m3ps**) in cubic meters per second.

Hydraulics

Figure 15-1. Flexible hydraulic transmission line.

```
template hose p1 p2 = arg1, arg2
hyd_mks  p1, p2         # Hydraulic connections
number arg1, arg2       # Declare arguments of template as before

{                       # Body start
 parameters {
    # Perform one-time parameter processing...
    }
 values {               # Implement governing equations...
    pressure = p_Npm2(p1) - p_Npm2(p2) # hydraulic operator analogous to v( )
    }
 equations {
    q_m3ps(p1->pmid) += q1_mks       # hydraulic operator analogous to i( )
    q_m3ps(pmid) += d_by_dt(vol_f)
     }
}                       # Body end
```

This is a simple model that possesses the effects of viscous losses and inertia capacitance due to fluid compressibility and line expansion. Effects that would be considered for a more complex implementation would be laminar and turbulent flow and cavitation. The equations for this model express the through variable (flow) as a function of the across variable (pressure). Note the use of the += operator. Because the compressibility effect is seen at each pin independently, branches cannot be used. The += operator is used to add this flow (the through variable) to each pin.

15.3 Example Design—Simple Hydraulic Press

Figure 15-2 shows a basic fluid power system. The system consists of an ideal fluid flow source supplying a spring-loaded accumulator. The pressure at the supply is limited by a relief valve that has a cracking pressure of 10 MPa, and is fully open at 12 MPa. The supply pressure is applied to the load through a 3-way valve and a length of flexible hose. The actuator extends the load under pressure, and has an internal return spring when the pressure is removed. The three-way valve alternately connects the supply or exhaust (shown as a "ground" symbol) pressure, and its spool position is controlled by an ideal mechanical position source.

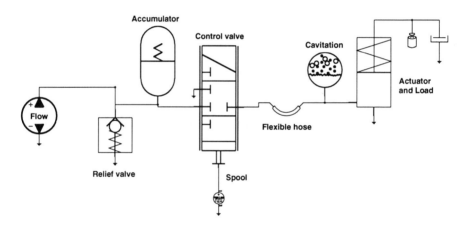

Figure 15-2. Simple hydraulic press.

Simulation results

Figure 15-3 shows the supply pressure and the accumulator fluid flow for a single operating cycle of the press. The supply is turned on at time equal to zero. The pressure increases steadily as the accumulator fills, due to compressing the internal spring of the accumulator. Note the net positive fluid flow rate of 40 μm³/s, into the accumulator. When the pressure exceeds 10 MPa, the relief valve opens and holds the pressure near the cracking pressure. The flow into the accumulator ends as pressure stabilizes. At 13 seconds, the three-way valve opens, and flow *from* the accumulator (negative direction) supplies the actuator. At 14 seconds,

Hydraulics

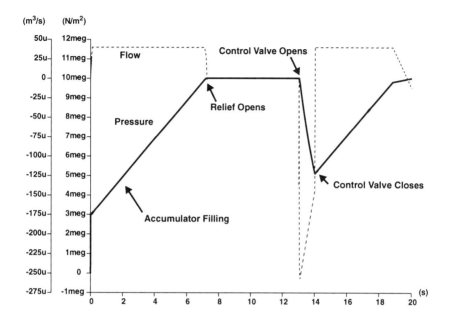

Figure 15-3. Pressure and flow into the accumulator.

the valve is again closed. The pressure recovers as the positive flow re-fills the accumulator.

Figure 15-4 shows the load position and the actuator pressure during this same interval. The pressure rises very rapidly when the valve is opened to supply. It subsequently falls to a lower steady level (not shown), once the load has accelerated to a constant velocity. The steady pressure reflects the force of the actuator return spring and the viscous damping of the moving load. Once the valve is closed, the pressure momentarily drops to a value below zero (i.e., below atmospheric pressure), as the load momentum causes cavitation. The actuator pressure returns to its previous level while the load transitions back to its initial position, and once the load bottoms, the pressure drops to zero.

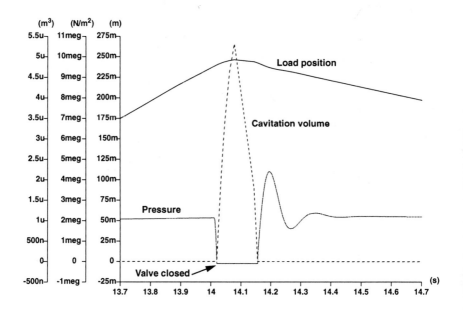

Figure 15-4. Load overrun at peak actuator pressure and cavitation volume.

15.4 References

1. H. E. Merritt, *Hydraulic Control Systems*, John Wiley & Sons Inc., 1967.
2. *Optional Template Library* manual, Ch. 36, Analogy, Inc., May 1994.

CHAPTER 16

Electro-thermal Systems

Due to the increasing current density and higher power requirements of advanced power semiconductor devices such as the IGBT (Insulated Gate Bipolar Transistor), these devices dissipate a considerable amount of heat. Regardless of the electrical quality of the power semiconductor devices, such devices will fail if the heat cannot be removed effectively. In addition, the electrical performance of the system can be limited by device heating and thermal coupling between devices. Therefore, concurrent simulation of the electrical and thermal aspects of power electronic systems is essential for effective computer-aided design of these systems. Electro-thermal simulations also enable the design of the thermal management systems (typically one of the most costly aspects of the overall system design) to be incorporated into the original design of the electronic system.

Typical power electronic system design considerations include control system analysis, gate level logic simulation, the interaction of power semiconductor devices and magnetic components with the external circuit, and the thermal management of the power dissipated in the semiconductor devices. However, traditional circuit simulation programs such as SPICE were developed for integrated circuit analysis and do not provide the capability to effectively analyze many of these aspects of power electronic systems. As a result, various simulation tools have been developed that specialize in the analysis of specific aspects of advanced electronic systems, such as power system simulation [1], thermal network analysis [2], and physics-based semiconductor [3] and magnetic component simulation. However, these separate tools require the designer to use different simulation environments, and only permit limited interaction between the various aspects of the overall system design.

The purpose of this chapter is to describe another technological area that an AHDL can help to address. The example shows how thermal management considerations can be readily incorporated into the overall design process for a pulsewidth-modulated (PWM) inverter. After describing some aspects of thermal modeling, a full-bridge, voltage-source PWM inverter will be presented as an example of thermal modeling within an electrical system.

16.1 Electro-thermal Network Modeling and Simulation

Although SPICE-based circuit simulators have traditionally simulated electronic circuits using detailed semiconductor device models, it is quite difficult to simulate the behavior of modern semiconductor devices that dissipate a considerable amount of heat. In the typical approach, a user specifies the temperature for the device models prior to simulation, which remains constant during simulation. This approach has become increasingly inadequate for modern power electronic devices and other advanced semiconductors because of their considerable power dissipation. This increases internal device temperature (self-heating) as well as the temperature of other adjacent semiconductors (thermal coupling).

The approach taken in the electro-thermal network simulation described in this chapter is to define the temperatures within the network as simulator system variables. This permits the simulator to solve for the dynamic temperature distribution within the thermal network in the same way that node voltages are solved for within an electrical network [4].

Alternative approaches for electro-thermal simulation have also been explored. Some representative techniques are described in [10, 11]. Many of these approaches combine different simulators to simulate the electrical and thermal portions of the system. SPICE is typically utilized as the electrical simulator. The thermal simulator falls into one of several categories. The technique in [10] is to use a finite element approach, while that of [11] employs a macromodel-based technique. While these approaches may have their merits, one of the main reasons that there is motivation to provide these "glued" approaches to electro-thermal simulation is the lack of an adequate means of describing the electrical and thermal elements to a single simulator. An AHDL addresses this problem directly, thus making electro-thermal simulation more straightforward. The fundamental elements that are needed will be described in this chapter.

Electro-thermal modeling and simulation

Figure 16-1 shows how electro-thermal models for semiconductor devices couple electrical and thermal networks. These semiconductor device models (e.g., IGBTs and power diodes) have electrical terminals that are connected to the electrical network and a thermal terminal that is connected to the thermal network. The thermal network is represented as an interconnection of thermal components that allow a system designer to readily interchange different thermal components and examine different configurations of the thermal network. These thermal models for power modules and heat sinks contain multiple terminals and account for the thermal coupling between the adjacent semiconductor devices.

Electro-thermal semiconductor models are implemented in MAST by expressing current flow into electrical nodes and power flow into thermal nodes in terms of the simulator system variables. Simulator system variables are the across variables of each domain—voltages across the electrical nodes and the temperatures across the thermal nodes. The through variables are current at electrical nodes and power at thermal nodes. The simulator solves the system of electro-thermal equations by iterating all system variables until the current into each electrical node sums to zero (Kirchhoff's current law, KCL) and the power into each thermal node sums to zero (energy conservation).

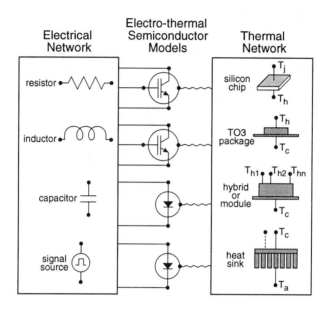

Figure 16-1. Connection of electrical and thermal networks through electro-thermal models for semiconductor devices (Reproduced with permission from National Institute of Standards and Technology).

In a network of thermal models, temperature is a system variable that is solved for by the simulator, as opposed to traditional circuit simulator semiconductor models in which temperature is a constant parameter specified by the user. Therefore, partial derivatives of the semiconductor model equations with respect to temperature would be required to implement dynamic electro-thermal effects into SPICE-based simulators, whereas the temperature effects are implemented into the Saber simulator by simply evaluating the temperature-dependent model equations in the MAST template.

Figure 16-2 shows an example of an electro-thermal network consisting of an IGBT electro-thermal model [5] and the thermal component models for the silicon chip, a TO247 package, and a TTC1406 heat sink [6]. The thermal network component models are connected to form a thermal network the same way electrical components form an electrical network.

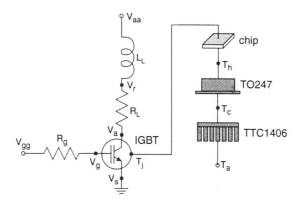

Figure 16-2. *Electro-thermal network for an IGBT with a TO247 package and TTC1406 heat sink (Reproduced with permission from National Institute of Standards and Technology).*

Electro-thermal semiconductor models

Figure 16-3 shows the structure of the electro-thermal semiconductor models indicating the interaction with the thermal and electrical networks through the respective electrical and thermal terminals [4]. These models use the instantaneous device temperature (temperature at the silicon chip surface T_j) to evaluate the temperature-dependent properties of silicon and the temperature-dependent model parameters. These values are then used by the physics-based semiconductor device model to describe the instantaneous electrical characteristics and instantaneous dissipated power. The dissipated power is calculated from the internal components of current, because a portion of the electrical power delivered to the device terminals is dissipative and the remainder charges the internal capacitances. The dissipated power calculated by the electrical model supplies heat to the surface of the silicon chip thermal model through the thermal terminal.

Implementing self-heating effects in a MAST template is straightforward. Since the Saber simulator does not require first derivative information, the focus remains on utilizing or

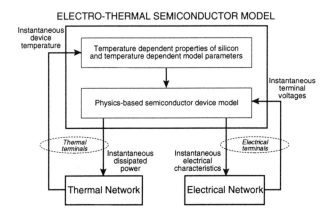

Figure 16-3. Structure of electro-thermal semiconductor device models (Reproduced with permission from National Institute of Standards and Technology).

developing the proper temperature-dependent expressions that are required for the device being modeled. As pointed out above, there are two fundamental types of relationships that must be accounted for. The first are the well-known temperature-dependent properties of silicon (or whatever material is being modeled). The second are those that describe the temperature dependence of the model parameters. These expressions may be simple or complex to derive depending on the nature of the model. The more physical the model, the easier that this tends to be. A highly empirical model will typically possess far more parameters, and characterization of the temperature dependence of the model may be very difficult.

There are only a few basic items that must be addressed to implement self-heating effects. The *pseudo*-template below outlines these items. These are described from top to bottom as the template is listed. First, a thermal pin is declared. This may be done with units of Kelvins, degrees Fahrenheit or degrees Celsius. While the units of degrees Celsius is the most common, typically the temperature will have to be converted to Kelvin internally for implementation purposes. However, the choice of units for the thermal pin is made based on those desired in reporting and plotting.

Electro-thermal Systems

```
template igbt_th a g k tj = arg1, arg2, ...# Template header
electrical a, g, k     # Electrical nodes
thermal_c tj           # Thermal node
number arg1, arg2, ...
{                      # Template body
# Local declarations
val tk tempj           # Junction temperature (Kelvins)
val ...                # Other local declarations
parameters {           # Parameters section (one-time calcs.)
                       #...Remove those items that are now functions of instantaneous
                       # temperature tempj and relocate them in the values section
    }
values {
    tempj = tc(tj) + 273.15    # Junction temperature in Kelvins
    #...Implement temperature-dependent properties of silicon (Table 1)
    #...Implement model parameter variation as a function of bias (as done in
    # static thermal case)
    #...Implement model parameter variation as a function of tempj_k
    #...Implement governing equations
    #...Calculate power based only on dissipative contributions
    pwrd = i1**2 * r1 + i2**2 * r2 + ...
    }
equations {
    #...Equations as implemented without self-heating...
    p(tj) -= pwrd              # Total power dissipation sourced out
                               # of thermal pin tj
    }
}
```

Next, the template arguments are declared, just as in other templates. The primary temperature that appears in all of the model relationships will be the instantaneous junction temperature. This entity is declared in the local declarations section of the template (**tempj** in this case). It is declared in Kelvins since many of the thermal properties of silicon are expressed in terms of Kelvins.

Other local declarations have to do with the components of power dissipation and those entities that were formerly calculated in the parameters section (for the static temperature case) that are now functions of the instantaneous junction temperature **tempj**.

If converting a static thermal model to a dynamic thermal model, the parameters section is actually simplified, since many one-time calculations now become iterative calculations that must appear in the values section.

In the values section, **tempj** is first defined and then converted to Kelvins if necessary. The temperature-dependent properties of silicon are then implemented. Next, the model parameters as a function of bias are calculated; then, these results are used to calculate the model parameters as a function of temperature. The governing equations of the model are implemented as before except that they now utilize the temperature-modified model parameters. Lastly, the dissipative portions of the governing equations are summed together to obtain a total power dissipation value. This value is subsequently "sourced" out of the thermal pin in the equations section, that being the only modification to that section.

Thermal network models

In a similar fashion to the procedure described above for adding thermal effects to an electrical model, purely thermal models can be created in MAST as well. In simulating an electro-thermal network, the thermal network is represented as an interconnection of thermal models. Each model represents an individual building block used by a designer to form the thermal network (see Figure 16-1). The thermal models are typically lumped approximations of the heat flow through the chip, package, or heat sink material. The order of the approximations determine the accuracy achieved.

A very accurate approach was developed by Hefner and Blackburn [6]. Their thermal models are parameterized in terms of structural and material parameters so that the details of the heat

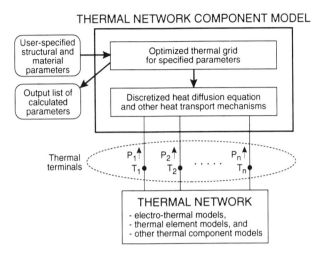

Figure 16-4. Diagram of the structure of the thermal component models (Reproduced with permission from National Institute of Standards and Technology).

transport physics are transparent to the user. However, the user can also specify the values of the structural and material parameters to provide accurate simulations for specific thermal models. Since these models are used in the example provided in Section 16.2, they are summarized here.

Figure 16-4 is a diagram of the structure of the thermal component models, indicating that these interact with the external thermal network through thermal terminals $T_1, T_2, ..., T_n$. The thermal component models are based upon a discretization of the heat diffusion equation for various three-dimensional coordinate system symmetry conditions and include the nonlinear thermal conductivity of silicon and the nonlinear convection heat transfer at the heat sink fins [6].

It is important to note that time constants for heat flow within the silicon chip, package, and heat sink are many orders of magnitude longer than the time constants of electronic devices and circuits. Because of this, self-heating effects behave dynamically— *even for circuit conditions that are considered to be steady-state for the electronic devices.* In addition, for circuit conditions that result in high power dissipation levels, the heat is applied rapidly to the chip and only diffuses a few micrometers into the chip surface. Therefore, the chip-heating process is nonquasi-static, so the temperature distribution within the thermal network depends on the rate at which the heat is dissipated.

16.2 Electro-Thermal Network for a PWM Inverter

Pulsewidth-modulated, voltage-source inverters are used extensively in power conversion, power conditioning, and motion control [7] due to their high-energy efficiencies. The design of a PWM inverter involves the analysis of the inverter topology, the control system, the logic circuits, the interaction of semiconductor devices with other circuit elements, the analysis of motors and other load devices, and the design of the thermal management system.

Circuit topology and models

The basic electro-thermal network used for the PWM inverter simulations in this chapter is shown in Figure 16-5 [8]. Although many variations of this basic inverter are used in practice, it is representative of the type of system analysis that is applicable to all inverter circuits.

The network of Figure 16-5 consists of switching control logic, IGBT switching devices and IGBT gate drivers in an H-bridge configuration, an electrical module of an equivalent motor load, and a thermal network for each IGBT. The control logic uses event-driven (digital) models to implement an open-loop, sine-triangle, pulsewidth-modulated control system for the switching devices. The right side of the network includes the electro-thermal models for the power semiconductor devices (IGBTs), as well as the equivalent electrical circuit for the motor load. The bottom of the network contains the thermal components for the semiconductor silicon chips, standard TO247 power semi-conductor packages, and TTC1406 heat sinks.

Input and control models

The logic gates and signal sources implement a behavioral representation of the open-loop, sine-triangle, pulsewidth-modulation control scheme. The input source (**vcntl**) provides a 60 Hz sinusoidal signal, which is compared with two triangle waves

Electro-thermal Systems **215**

to produce a pair of pulsewidth-modulated digital output signals. These signals are then clocked into the gate drivers of IGBTs 1 and 3. The clock signals are applied directly to the gate drivers of IGBTs 2 and 4. The models used in the control circuitry for this example contain both analog and digital signals.

Switching device models

The models for the IGBT gate drivers are also behavioral models that convert the digital inputs to analog IGBT gate drive signals. Here, the gate drive signals switch from 0 to 20 V with 50 Ω of series gate resistance. In this particular control scheme, the bottom IGBTs are turned on continuously during one half cycle of the inverter (60 Hz cycle), while the upper half (the opposite phase of the H-bridge) is switched at the clock frequency of the triangle waves with a varying duty cycle.

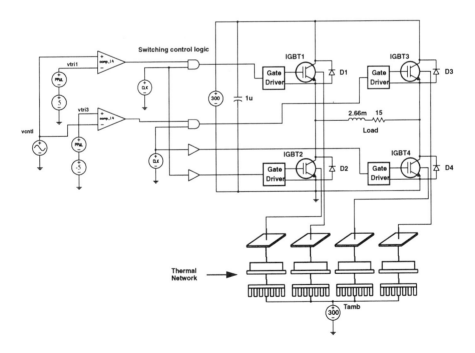

Figure 16-5. The basic electro-thermal network used for the PWM inverter simulations.

There are several levels of model representations that could be employed for the IGBT gate drivers. The drivers could be represented as discrete electrical components (i.e., transistor level), which would be useful for studying the interactions between the driver and the IGBT input characteristics. Transistor level models would also be helpful when designing new driver circuitry in order to understand how variations in the gate driver affect the IGBT waveforms or for designing snubber circuitry.

The inverter uses IGBTs as the switching devices (although models for other power devices such as MOSFETs or GTOs could be substituted) with anti-parallel connected power diodes. The model of the IGBT is the actual implementation based on the *pseudo*-template of the last section. It is a physics-based model capable of predicting switching losses within the device as well as the temperature rise associated with the dissipated power. It is also possible to substitute simpler, more computer efficient models for the devices for simulations in which the details of the switching devices are of little concern, such as when designing the control circuitry or simulating a larger system. Here, the detailed physics-based models are required to determine power dissipation and the influence of temperature on the device characteristics and circuit operation.

The motor load being driven by this inverter is represented here as an inductor with series resistance. However, a more elaborate motor model based on the methods described in Chapter 14 could be used.

Thermal network models

The thermal network for each IGBT in Figure 16-5 consists of an interconnection of thermal models for the following components:

- silicon chip
- TO247 package
- TTC1406 heat sink

These models describe the static and dynamic thermal impedance of each thermal component for the full range of power dissipation levels. Using the thermal network component models, a system designer can readily examine the effects of different thermal network topologies simply by changing connection points of the thermal terminals within the thermal network. For example, the behavior of the network in Figure 16-5 including thermal coupling between adjacent IGBTs and/or power diodes mounted on the same heat sink can be compared with the behavior of the same system, but with each power semiconductor device having a separate heat sink.

By connecting a temperature source to the thermal terminal (T_j) of the electro-thermal semiconductors, the electro-thermal network reduces to the traditional constant temperature circuit simulations. However, even this simplest of thermal models goes beyond the traditional SPICE-based models in which a single temperature must be specified for the entire electrical system.

Simulation results

Figure 16-6 shows simulation results for the motor load current and the input signals to the IGBT gates. The transient analysis was run at a relatively low switching frequency (900 Hz) to illustrate the behavior of the circuit. The duty cycle of the input signal to the IGBT gate drivers is varied using the sine-triangle comparison technique to produce a 60 Hz sinusoidal variation of the motor load current.

During the positive voltage phase of the 60 Hz inverter reference signal (e.g., between 16.7 ms and 25.0 ms in Figure 16-6), the gate control signal of IGBT1 is switched at the 900 Hz triangle wave frequency, while IGBT4 remains on. Also, during this phase, IGBT2 and IGBT3 remain off. Inverter operation during the negative voltage phase of the 60 Hz reference sine wave is similar to that during the positive voltage phase, except that the opposite phase of the bridge is switched on and off (i.e., IGBTs 2 and 3).

Figure 16-7 shows the results of simulating at a more realistic switching frequency of 20 kHz. PWM inverters are typically operated in this range in order to push the switching frequency beyond audible frequencies. However, considerably more heat is dissipated in the IGBTs due to the on-state losses for the large motor current and switching losses at this frequency.

Figure 16-7a shows the load current; Figure 16-7b shows the temperature waveforms at the silicon chip surface (T_{j1}), the chip-package interface (T_{h1}), and the package-heat sink interface (T_{c1}) for IGBT1. Notice that the silicon chip surface temperature waveform of IGBT1 appears to have spikes at a 20 kHz rate during the phase that the device is switching (e.g., 16.7 ms through 25.0 ms) due to the switching energy losses.

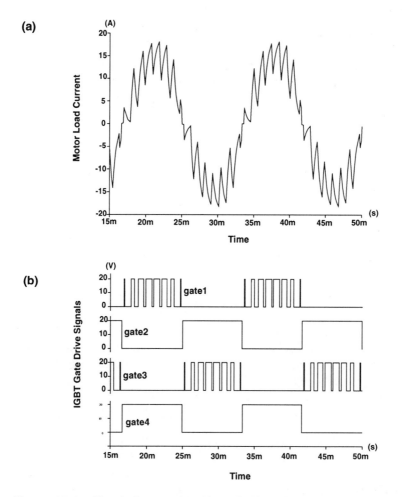

Figure 16-6. Simulations at 900 Hz switching a) motor load current and b) gate signals to the IGBTs.

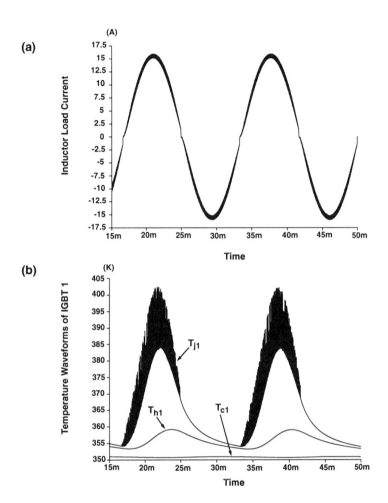

Figure 16-7. Simulations at 20.1 kHz switching a) load current, b) temperature waveforms at the silicon chip surface (T_j), the chip-package interface (T_h), and the package-heat sink interface (T_c) for IGBT 1.

Also notice that the chip surface temperature cools after the peak in load current (21 ms) and during the phase in which the device is off (e.g., 25 ms through 33 ms). The 60 Hz variations of the chip surface temperature waveform are due to the sinusoidal load current variation and due to cooling during the half of the 60 Hz cycle in which the device is off all of the time.

More simulation results are available in [8]. Essentially, they indicate how taking into account the thermal effects of this system influences the overall electrical efficiency of the inverter by observing the difference in the switching energy of the IGBT devices with and without these effects.

From the temperature waveform of Figure 16-7b, it is evident that the thermal response of the silicon chip determines the IGBT temperature variations during the device 20 kHz switching cycle (from 385 K to 405 K). This is because the temperature at the package header (T_h) does not change during the 20 kHz cycles. The thermal response of the silicon chip and TO247 package determines device temperature variations during a single phase of the 60 Hz sinusoidal output (from 355 K and 385 K), because the temperature at the package case (T_c) does not change. For the relatively high frequencies (with respect to the time constants of the thermal network) of the electrical network, the temperatures within the heat sink do not change substantially. However, the heat sink does influence the thermal response to changes in the power dissipation level over longer periods of time, such as during the system start-up or during load impedance changes.

Conclusions

This chapter illustrates how an AHDL can be used to create models for electro-thermal analysis. The example of a PWM inverter provided a detailed demonstration of combining electrical (including analog, digital, microelectronics and power electronics) and thermal models into a single system simulation. The full electro-thermal simulations in the example are required to accurately predict the total energy loss and the overall efficiency of the inverter network of Figure 16-5. The average power dissipation in IGBT1 calculated including the device self-heating is 20.5 W, whereas the value calculated assuming a constant temperature of 300 K is much less, 15.3 W. The overall efficiency for the PWM inverter can be calculated by integrating the power dissipated in the load resistor over a complete 60 Hz inverter cycle, and dividing by the integral of the power out of the 300 V power supply. The overall energy efficiency calculated for the simulation including the self-heating effects is 96.1%, whereas the constant temperature simulations effectively neglect 15% of the total energy loss in the inverter. Because the simulations including self-heating accurately describe the dissipated power, the simulations can be used to select an IGBT type that minimizes the energy losses and to design a thermal management system that is suitable for the calculated power dissipation levels.

16.3 References

1. Boeing Company, *EASY5 Version 3.2 New Features Descriptions*, 1985.
2. G. N. Ellison, *Thermal Computations for Electronic Equipment*, Van Nostrand Reinhold, New York, p. 218, 1984.
3. A. R. Hefner, "Semiconductor Measurement Technology: INSTANT —IGBT Network Simulation and Transient Analysis Tool," NIST Special Publication 400-88, 1992.
4. A. R. Hefner and D. L. Blackburn, "Simulating the Dynamic Electro-Thermal Behavior of Power Electronic Circuits and Systems," *IEEE Trans. Power Electronics*, pp. 376-385, Oct. 1993.
5. A. R. Hefner, "A Dynamic Electro--Thermal Model for the IGBT," *IEEE IAS Conf. Rec.*, p. 1094, 1992.
6. A. R. Hefner and D. L. Blackburn, "Thermal Component Models for Electro-Thermal Network Simulation," *Proc. IEEE Semiconductor Thermal Measurement and Management SEMI-THERM Symposium*, p. 88, 1993.
7. IEEE Spectrum: Kilowatts on Order, p. 32, Feb. 1993.
8. H. A. Mantooth and A. R. Hefner, "Electro-thermal simulation of an IGBT PWM inverter," *IEEE PESC Conf. Rec.*, pp. 75-84, June 1993.
9. E. D. Wolley and W. R. Van Dell, "Fast Recovery Epitaxial Diodes (FRED's)," *IEEE IAS Conf. Proc.*, p. 655, 1988.
10. W. Van Petegem, B. Geeraerts, W. Sansen and B. Graindourze, "Electrothermal simulation and design of integrated circuits," *IEEE Journal of Solid-State Circuits*, Vol. 29, pp. 143-146, Feb. 1994.
11. S. S. Lee and D. J. Allstot, "Electro-thermal simulation of integrated circuits," *IEEE Journal of Solid-State Circuits*, Vol. 28, pp. 1283-1293, Dec. 1993.

CHAPTER 17

Magnetics

17.1 Overview

The study of magnetic properties has many similarities to that of electrical properties. Because magnetic behavior can be expressed as ordinary differential equations for continuous time simulation, the same types of analyses may be performed on magnetic models as those that are done on electrical models. Consequently, the MAST AHDL allows straightforward implementation of magnetic templates, using the same constructs and procedures for electrical templates introduced in Chapters 1 through 12.

A magnetic field may result from electrons flowing in a wire, or it may exist independently as a permanent magnet. Regardless of its origin, a magnetic field emanates lines of flux. These lines of flux exist in continuous loops in a magnetic path, radiating outward from one end of the magnet to the other. These ends are designated as the north and south poles for reference labels of the magnetic circuit, much like the plus and minus labels of an electrical circuit. Furthermore, the driving force for these lines of flux is known as the magnetomotive force (*mmf*) across the path. Thus, the magnetic through and across variables are flux and *mmf*, respectively.

As with models for other non-electrical technologies (such as mechanical and hydraulics), once the through and across variables have been determined, the unit and pin definitions for magnetic templates can be implemented accordingly. This allows full description of the governing equations for the technology.

Electromagnetics

The area of electromagnetics focuses primarily on the electrical functions and applications that can be realized from magnetic material properties, rather than on the elaborate physics that underlie the achievement of material parameters. Such applications illustrate the meaning of basic magnetic properties and parameters to the extent that they influence the behavior of an electrical circuit. For this reason, magnetic modeling remains within the realm of electrical devices using magnetic cores.

Electromagnetic devices exhibit the interaction of electrical and magnetic characteristics by winding conductive wire around a core with magnetic properties. Because the MAST modeling language can model both electrical and magnetic behavior separately, it is also possible to model electromagnetic interaction. This is done by modeling the magnetic properties of the core while an electrical stimulus is being applied to the wire coiled around it. Further, the magnetic behavior of a core material can be highly nonlinear with changing electrical stimulus.

This magnetic-electrical interaction is as follows:

1. An electron flowing through a wire induces a magnetic field. The characteristics of this field are principally related to the amount of current flowing in the wire, along with the shape into which the wire has been coiled.

2. A magnetic field in proximity to a wire creates a flow of electrons in the wire.

An electromagnetic device (such as an inductor or transformer) is of particular interest because it incorporates both types of interaction. This requires that the device accurately model the magnetic and electrical interactions.

The inductor

The simplest electromagnetic device is the inductor. Because it contains both electrical and magnetic circuits, the inductor provides an opportunity to investigate their interaction.

An inductor consists of a conductive wire (or *winding*) that has been wound into a coil around a magnetic *core*. The coiled shape of the winding maximizes the magnetic field that is produced by the flow of electrons. This magnetic field is said to exist within the inductor's core—the volume around which the inductor is coiled. The core may consist of either a solid magnetic material (such as iron or an iron alloy), or free space, or a combination of both. Regardless of the material, it is within and around the winding and core that the magnetic "circuitry" exists. Thus, the inductor model needs to contain two models from different technologies:

- a winding, which serves as a path for electrical quantities
- a core, which serves as a path for magnetic quantities

When the inductor is connected to an electrical circuit such that a current flows in the winding, it begins the following interaction:

- An *mmf* (magnetomotive force) is generated in a magnetic path by a current passing through a coil of wire wrapped around the magnetic path.
- In turn, flux flowing in a magnetic path can generate an emf (electromotive force, or voltage) in a coil of wire wrapped around the magnetic path.

17.2 Fundamental Magnetic Quantities

The magnetic circuit occurs within the core of an electromagnetic device. The magnetic field that emanates from the core is characterized by the following quantities.

NOTE

Several sets of units exist for the measurement of magnetic properties, the most common of which are the MKS (or SI) set and the CGS (or Gaussian) set. Many magnetic calculations depend upon which set of units has been selected. In addition, there are methods for converting between sets of units.

Magnetomotive force (mmf)

In an electrical circuit, the force applied across an electrical path to produce flow of charge (current) is called electromotive force or voltage (v).

In a magnetic circuit, the force applied across a magnetic path to produce flux is called magnetomotive force (mmf). This causes flux lines to be set up within the magnetic material.

The magnetomotive force results from the interaction of the winding and the core. It is proportional to the product of the number of winding turns (N) around the core and current through the winding (i). The equation for *mmf* is:

$$mmf = N \cdot i \qquad (17\text{-}1)$$

The unit for *mmf* is ampere-turn (A-t), which follows from the product of number of turns and current. Note that this quantity depends on a quantity from the electrical circuit, current.

"Ohm's Law" for magnetic circuits

For an electrical circuit, the relationship between current, voltage, and resistance (I = V/R) is known as Ohm's Law. For a magnetic circuit, a similar relationship exists between flux (Φ), magnetomotive force (*mmf*), and reluctance ($\mathfrak{R}$):

$$\Phi = mmf/\mathfrak{R} = (N \cdot i)/\mathfrak{R} \qquad (17\text{-}2)$$

Hence, *mmf* is the across variable and flux is the through variable for magnetic circuits. Because *mmf* is defined as the current flowing through turns of a winding, an increase in either the number of turns or the amount of current will increase the amount of magnetic flux.

Flux (Φ)

As stated earlier, a magnet radiates flux lines from its north pole to its south pole (Figure 17-1). Because all magnetic lines of flux must form a closed loop, these lines then return to the north pole *through the length of the magnet.*

Flux lines do not have origins or terminating points—they exist in continuous loops. Magnetic flux, the through variable, is established in the core as a result of magnetomotive force, the across variable. It is important to note that, although flux is the through variable at a given magnetic pin, it is not a "flowing" quantity of discrete entities like electrical current. The symbol for flux is Φ (as a template parameter, its abbreviation is ƒ) and it is measured in webers (Wb).

Within the core, flux lines are spaced equally. Outside the core, they are symmetrically distributed. Any given continuous line of magnetic flux will attempt to occupy as small an area as possible. The strength of a magnetic field in a given region is directly related to the density of flux lines in that region. *In other words, the more flux lines that occupy a region, the greater is its flux density, B.*

When a magnetic field results from electrons flowing through a wire, the direction of magnetic flux is found by what is known as

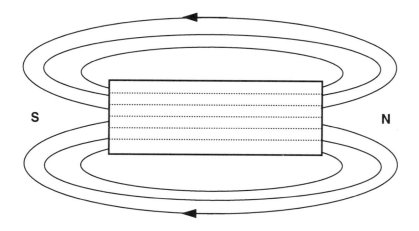

Figure 17-1. Magnetic lines of flux.

the "right hand rule." That is, the magnetic flux around the wire is oriented in the same direction as the fingers of the right hand are pointing when the right thumb points in the direction of conventional current. If the wire is wound in a single turn coil, the resulting flux will flow in a common direction through the center of the coil. A coil of more than one turn produces a magnetic field that exists in a continuous path through and around the coil.

Flux density (B)

The number of flux lines per unit area of the core is called the *flux density* (B) and is measured in tesla (T) or gauss (G) units. Flux density is a vector quantity that is equal to the surface integral of Φ over the core area:

$$B = \int \Phi \, ds \qquad (17\text{-}3)$$

When Φ is constant across the surface, this simplifies to the following:

$$B = \Phi/A \qquad (17\text{-}4)$$

where (in MKS units):

B = flux density (T)

Φ = magnetic flux (Wb)

A = area of core (m^2)

Using this system of units, $1\,\text{T} = 1\,\text{Wb/m}^2$.

Permeability (μ)

If two cores have the same physical dimensions but are made of different materials, then magnetic strengths will differ accordingly. This difference is reflected by a greater or lesser number of flux lines established in the core. The measure of the ease with which magnetic flux lines can be established in a material is called the *permeability* (μ). The higher the value of permeability, the more flux lines result from a given value of current flowing in the winding around the core. It is measured in webers per ampere-meter (Wb/A-m) or, alternately, in henrys per meter (H/m).

Permeability is analogous to conductivity in electrical circuits. The permeability of free space (vacuum) is labeled μ_0. For all practical purposes, the permeability of nonmagnetic materials is regarded as the same as free space. The permeability of free space is:

$$\mu_0 = 4\pi \times 10^{-7} \ Wb/A\text{-}m = 1 \ G/Oe \tag{17-5}$$

Magnetic field strength (H)

The magnetomotive force (mmf) per unit length of the magnetic path (l) is called the magnetic field strength (H). Magnetic field strength is a vector quantity given by the equation:

$$H = mmf/l = (N \cdot i)/l \tag{17-6}$$

The unit for H is amperes per meter (A/m) or oersted (Oe), depending on the set of units selected.

Note that H can also be expressed as the product of turns and current per unit length, which is generally a more useful way of expressing it. H is independent of the type of core material; it is determined solely by the number of turns (N), the current flowing through the winding (i), and the length of the magnetic path (l).

The applied H has a pronounced effect on the resulting permeability (μ) of the core material; as H increases, μ rises to a maximum, then drops to a minimum.

17.3 Nonlinear Magnetics

As indicated in the previous section, there is an important relationship between the flux density (B) and magnetic field strength (H) in the core material. In all probability, B will be related to H in a nonlinear way.

The curve representing this relationship is known as the B versus H curve, or simply the B-H curve. Figure 17-2 shows a typical B-H curve, with B on the vertical axis and H on the horizontal axis. This curve indicates the response of B as H varies monotonically—first to a maximum positive value (the lower curve), and then to a minimum negative value (the upper curve).

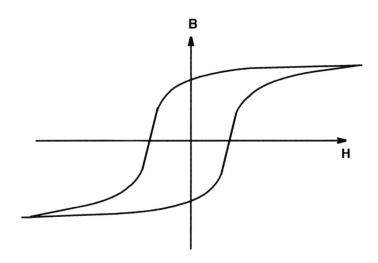

Figure 17-2. Typical B-H curve.

As indicated by the two curves, the value of B for a given value of H is different when H is increasing than when H is decreasing. This phenomenon is known as *hysteresis*. It is a property of virtually all magnetic materials. Note that neither the upper nor the lower curve passes through the origin. The origin is contained in the "envelope" between the two curves. The steepness of the B-H curve is determined by the value of permeability (μ) of the core material. For nonlinear magnetics:

$\mu = dB/dH$ (17-7)

The MAST AHDL allows the creation of models that provide nonlinear characterization of magnetic materials and electromagnetic devices.

Winding model

The following *pseudo*-template (**wndg**) demonstrates a rudimentary model for the winding of an inductor or a transformer. It has four connections, two of which are declared as *electrical* pins (**ep, em**) and two of which are declared as *magnetic* pins (**mp, mm**). This template serves as the interconnecting element between electrical and magnetic circuits, containing both magnetic and electrical models. Because of this, *the winding template is the key* *to creating electromagnetic models for simulation.* When used with a model for a nonlinear core (which incorporates the hysteresis of the B-H curve), a winding template links the nonlinear magnetic properties of the core to the rest of the electrical circuit [1].

```
template wndg ep em mp mm = n, r
electrical  ep, em               # Electrical pins
magnetic    mp, mm               # Magnetic pins
number      n,                   # Number of turns in winding
            r=0,                 # Winding resistance
{
var f       f                    # Flux through winding
var i       i                    # Current through winding
var nu      didt, dfdt
val mmf     mmf                  # mmf drop across winding
val v       v,                   # Total voltage drop across winding
            vdrop                # Voltage drop due to winding resistance
val l       l
number      scalefact=1meg       # Scale factor for numeric stability
parameters{
    # Perform one-time parameter processing...
    }
values {   # Implement governing equations...
           mmf1 = mmf(mp) - mmf(mm) # magnetic operator analogous to v( )
           }
equations {
    i(ep->em) += i
    f(mp->mm) -= f
    i: v = d_by_dt(f*n) + vdrop
    f: mmf = i*n
    didt: didt = d_by_dt(i/scalefact)
    dfdt: dfdt = d_by_dt(f*n/scalefact)
    }
}
```

17.4 Power Supply With Electromagnetic Models

Figure 17-3 shows an elementary power supply design containing a two-winding transformer and a pair of inductors acting as filter chokes. Although somewhat antiquated, this type of power supply design demonstrates the effects of nonlinear magnetics particularly well. The magnetic characteristics of the transformers and inductors are subject to changes in operating conditions, which can affect circuit performance. The overriding concern is to determine if the magnetics are saturating under operating conditions (i.e., if an increase in winding current produces no increase in flux density of the core). A linear magnetic model can be used to perform a quick check to determine approximate saturation levels. If they are near saturation, the design can be altered appropriately to operate below those levels so that saturation does not occur.

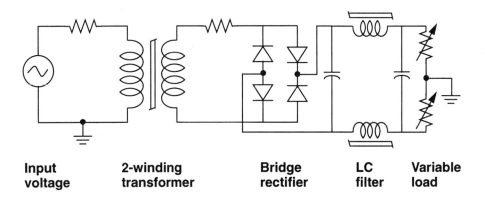

Figure 17-3. Power supply with electromagnetic models.

The power supply shown in Figure 17-3 consists of the following stages:
- sinusoidal input voltage
- two-winding step-down transformer (100:1 turns ratio)
- full-wave bridge rectifier

- LC filter for each half wave of rectification
- variable resistive load

The output is intended to deliver 200 mA at ±125 VDC. The variable resistors have been added to model a varying load that can be unbalanced. The four diodes are actual rectifier models (not ideal). The transformer and inductors also represent realistic models, including winding resistance and nonlinear B-H core characteristics.

The simulation consists of applying power to the input and sweeping the values of the load resistors to draw increasingly more current. This is done first with linear core models for the transformer and chokes; it is then repeated using corresponding nonlinear core models. The principal effect to be observed is the response of load current and voltage for the linear and nonlinear models, comparing the extent to which ripple and saturation are shown. In addition, the B-H characteristic of the transformer core can be determined from its nonlinear model.

Simulation results

Figure 17-4 shows the output voltage as a function of an increasing load current (which results from a decreasing load resistance) using linear and nonlinear magnetic models for the transformer and choke inductors. In Figure 17-4a, note the marked overshoot at turn-on, resulting from the fact that there are no losses to damp out the transients.

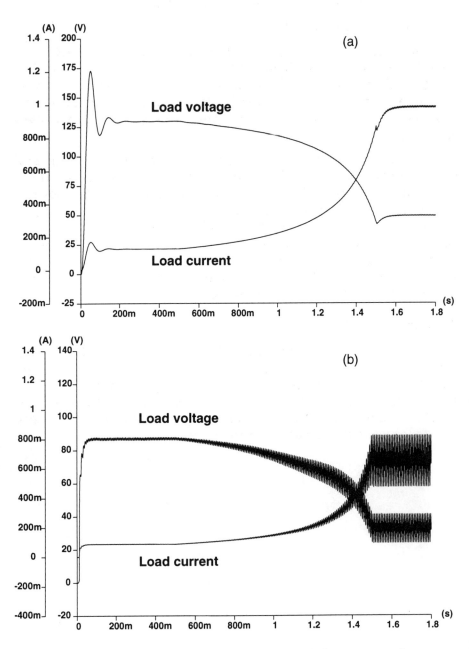

Figure 17-4. Output voltage and currents (a) with linear magnetic models (b) with nonlinear magnetic models.

Figure 17-4b shows that the front corner of the turn-on transient is fairly well damped (due to hysteresis and core losses). The output stabilizes to its normal level within 100 ms of turn-on. As the load current increases, so does the ripple amplitude. When the load current exceeds 750 mA, the ripple amplitude becomes unacceptably large. The increase in ripple results from the core saturation of each choke. This reduces its inductance and thus its filtering capability. At the same time, the output voltage has decreased to less than 100 V, well below its rated output. The drop in output voltage results primarily from the resistance of the choke and transformer windings.

As is readily apparent, one of the most obvious differences between the two graphs is the marked overshoot at turn-on when using linear models (Figure 17-4a). The output level of the linear models is a few volts higher than the supply with the nonlinear models. This difference is attributable to the two main factors:

- The core losses present with the nonlinear transformer and choke models.
- Some of the output voltage of the supply with nonlinear magnetics is contained in higher harmonics that are generated by the nonlinear conversion of the input sinusoid. These harmonics are attenuated by the output filter, but not before they create heat in the transformer (I^2R losses).

Figure 17-5 shows the hysteresis behavior of the magnetic material used in the core of the transformer. This represents the magnetization of the core in response to the supply going from no load to full load. The horizontal axis (H) is the magnetizing force, which is directly proportional to winding current. The vertical axis (B) is the flux density of the core material, which increases along a different curve than when it decreases.

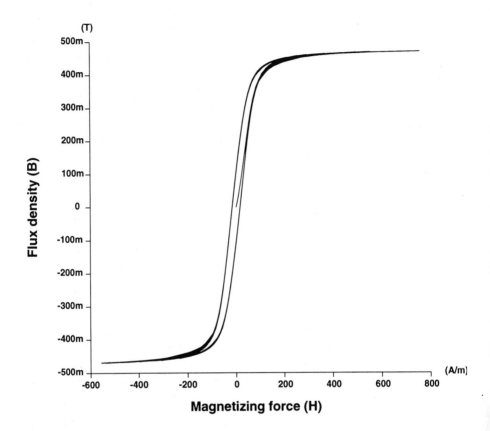

Figure 17-5. B-H hysteresis modeled by transformer core.

17.5 References

1. R. Buntenbach, *Analogs Between Magnetic and Electrical Circuits*, Electronic Products, pp. 108-113, Oct. 1969.
2. *Power Supply With Non-Linear Magnetics*, Pub. MA-0118, Analogy, Inc., Jan. 1993.

CHAPTER 18

Summary

This book is intended to provide a basic understanding to both practicing engineers and college students of what is needed to develop simulation models with an analog hardware description language (AHDL). Using the MAST modeling language from Analogy, Inc., as a specific implementation, the attempt has been to demonstrate how modeling with an AHDL can make computer simulation more efficient, powerful, and valuable. By way of explanation and example, the general terms of what modeling with an AHDL is all about have been spelled out using MAST and the Saber simulator to develop that understanding.

As with any publication, this book has had to endure the omission of topics because of time and space constraints. In some cases, interesting details regarding the models that were, or could have been, developed within an area of study or technology were left out in order to stay focused on the more immediate matters of using a modeling language. However, the references in this book indicate that a large volume of in-depth modeling has been implemented in MAST [1-10]. Some chapters used real applications from published papers as the starting point for addressing the detailed aspects of a particular modeling topic, while other chapters were created to provide simple demonstrations.

The intention has been to make clear that the MAST language can be used to create models for such diverse applications as transmission lines, optical systems, jet engines, neural networks, or just about any area of scientific inquiry for which differential equations can be formulated. Exploration of these areas and others like them indicate the true potential that modeling with an AHDL will realize in the years to come.

18.1 References

1. C. L. Ma and P. O. Lauritzen, "A simple power diode model with forward and reverse recovery," *IEEE Trans. Power Electronics*, vol. 8, pp. 342-346, Oct. 1993.

2. A. R. Hefner and D. M. Diebolt, "An experimentally verified IGBT model implemented in the Saber circuit simulator," *IEEE PESC Conf. Rec.*, p. 10, June 1991.

3. G. A. Franz, "Multilevel simulation of power converters," *Proc. IEEE Applied Power Electronics Conf.*, pp. 629-633, Mar. 1990.

4. J. M. Donnelly, C. Siegel, and D. Witt, "Design analysis of an electronically-controlled hydraulic braking system using the Saber simulator," *Proc. International Congress & Expo.*, pp. 1-5, Feb. 1994.

5. R. S. Cooper, "Rapid prototyping of an advanced motion controller," *Proc. AAS Guidance and Control Conf.*, 17 pp, Feb. 1992.

6. T. Koskinen and P. Y. K. Cheung, "Hierarchical tolerance analysis using behavioural models," *IEEE Proc. CICC*, pp. 3.4.1-3.4.4, May 1992.

7. R. Yacamini, L. Hu, and I. D. Stewart, "Electric drives on ships and oil platforms," *Inst. of Marine Engineers*, 14 pp., Feb. 1992.

8. R. Farrington, M. M. Jovanovic, and F. C. Lee, "Modeling losses and stresses of high frequency power converters using Saber," *Proc. IEEE Applied Power Electronics Conf.*, 5 pp., Mar. 1991.

9. J. E. C. Brown, M. Alexander, and D. F. Bowers, "Mixed-mode simulation of a continuous-time $\Sigma\Delta$ ADC," *IEEE Proc. Int. Symp. on Circuits Syst.*, pp. 1915-1918, May 1990.

10. N. T. Carnevale, T. B. Woolf, and G. M. Shepherd, "Neuron simulations with Saber," *J. Neuroscience Methods*, vol. 33, pp. 135-148, 1990.

APPENDIX A

Reference Information

This appendix covers some of the basic syntax, operators, and constructs used in MAST templates, along with some information on newton steps, which some types of models require for simulation.

A.1 MAST Reference

General syntax

The MAST modeling language is a high-level language that has syntax rules somewhat like the C programming language. It is block-structured, which means that logical programming blocks start with a left brace ({) and end with a right brace (}). Unlike other high level languages like C and Pascal, MAST statements do not have to end with a semicolon (;), although the semicolon is a valid end-of-statement character.

Any line ending with one of the *line-continuation* characters listed below is assumed to continue on the next line. Otherwise, the line is assumed to terminate.

\ + - / * & | < > ([{ = ,

The MAST language parser ignores blank lines, the pound sign (#), and any characters following a pound sign to the end of that line. A pound sign can start anywhere within a line.

Identifiers are names for variables, templates, netlist nodes, reference designators, etc. They must start with an alphabetic character or an underscore (_), followed by any mixture of alphabetic characters, digits, and underscores. An identifier may have any number of characters, all of which are significant. MAST is case-insensitive.

Connection points (pins)

There are several different kinds of MAST models that the Saber simulator can use in a system:

- Continuous analog (physical systems, such as electrical)
- Event-driven (such as digital)
- Data flow (such as control)

Typically, each of these categories has its own type of connection point (for example, an event-driven model uses state connection points). This communicates the characteristics of that type of model to the rest of the design. Note that it is possible to combine model types within a template, thus a template can have different types of connection points (digital inputs and analog outputs, for example).

NOTE

In general use, the terms "pins" and "connection points" are often used interchangeably. In the MAST language, a pin is a specific type of connection point that uses through and across variables (i.e., not all connection points are pins).

Include files

You can use the < character followed by a file name to make the contents of that file available to a template (i.e., an include file). An include file can contain any information that is part of a template, including a complete, stand-alone template. The < character must be the first character on the line calling the include file, and the include file must reside in the data search path of the simulator. For example, one could create a file called **constant.sin** that declares number type parameters and assigns constant values to them (such as **pi=3.141592653589793**). You would then include this file in a template as follows:

<constant.sin

Unit definitions

> **NOTE**
>
> *Analogy products contain an include file (named* **units.sin***) that provides unit definitions for all templates used with the Saber simulator. This file is automatically loaded upon invocation. The following information is provided for reference only—it is almost never required when writing a template.*

The MAST language allows one to define units for any given type of template variable, including connection points. Units can be made part of the definition of analog pins (for post-processing display), or they can be part of the declaration of state (digital) connection points.

In a template, unit definitions must be declared before the declaration of any connection points that use them. However, it is generally more convenient to define units for all models in a separate file and include that file automatically with invocation of the simulator.

For example, the following declarations would be used for the through and across variables associated with *analog* electrical pins:

        unit {"A", "Ampere", "Current"} i
        unit {"V", "Volt", "Voltage"} v

Further, the following declaration would define a set of units (named **logic_4**) for four logic states associated with *digital* connection points:

unit state { l4_0, "0", "0", "low.1",
l4_1, "1", "1", "high.1",
l4_x, "x", "x", "middle.1",
l4_z, "x", "z", "middle.1"} = logic_4 = l4_x

Pin definitions

Pin-type connection points need to be defined in terms of the units they will use for their associated through and across variables. This is done in conjunction with the unit definitions (above) for those variables, generally in a separate file that is automatically included with invocation of the simulator. These variables are then applied globally to all models. For example, the following definition for electrical pins uses the unit definitions for through and across variables, current and voltage:

pin electrical through i across v

Any template that declares a connection point to be **electrical** will automatically use the units for current and voltage as they are defined in the global include file.

Parameters

A parameter is a template variable that is assigned a constant value for a given simulation analysis. There are two general categories of parameters that determine where this value is assigned:

> *Argument* parameter—specified by the template user at the design (netlist) level. An argument can be changed with an alter command during a simulation session, but it cannot be changed in the parameters section of the template.
>
> *Local* parameter—specified by the template writer. A user does not have the ability to specify the value of a local parameter at simulation run time (i.e., the template must be edited to change a local parameter value).

Parameter data types

Each parameter in a template (including arguments) must be declared as a particular *type*. Each type can be considered as either simple or composite, although those terms have no meaning in MAST. A simple parameter is one that can accept only one value assigned to it. A composite parameter can consist of multiple simple parameters or even other composite parameters.

number	A simple parameter that can assume the value of any integer or real number.
enum	A simple parameter that can assume only one word from a set of alphanumeric (enumerated) words. This set is defined in the template as a comma-separated list between braces within the declaration.
string	A simple parameter that can assume any string constants of zero or more alphanumeric characters enclosed between quotation marks. Unlike an **enum**, the values for a string constant do not need to be defined in the declaration.
struc	A composite parameter that is used to group multiple parameters (that may or may not be of the same type) together under a single name for a collective purpose. This provides a convenient *structure* for working with a large number of related parameters, such as the basic model parameters of a diode (although a structure is not limited to grouping simple parameters).
union	A composite parameter that declares multiple parameters (that may or may not be of the same type) under a single name, but only allows one of them to be active. This provides a "choice" function so that the selection of any parameter within the union can be made conditional.

Single- and multi-dimensional arrays are also available; however, arrays are not explicitly declared as are other parameter types. An array is specified as an ordered collection of one of the other types, such as an array of numbers or structures.

Table A-1. Parameter data types.

Data type	Name	Examples
Simple	**number**	**number res=47k**
	enum	**enum {yes, no} rundc=no**
	string	**string matl="3c8"**

Composite	**struc**	**struc {**
		   number v=undef,
		               i=undef
		} vi_pairs=()

	union	**union {**
		   number v=5,
		            i=3m
		   struc {
		      number mag=undef,
		                    ph=0
		   } ac=()
		} total=()

Array	—	**number tc[2] = [200u,30u]**

Expressions

An expression is an operation performed on one or more parameters or variables. This is done using the unary or binary operators described below. An expression can contain multiple operations separated by parentheses. The result of an expression can be assigned to another variable (see below). The following are examples of expressions:

   out1 + out2
   2*pi
   input/(2*pi)

Unary and binary operators

Operators are special characters that perform mathematical functions in an expression (except //, which is used to concatenate strings). These operators are listed below in order of precedence:

~	(boolean not)
**	(exponentiation)
* /	(multiplication, division)
+ -	(addition, subtraction)
//	(string concatenation)
< <= > >=	(less than, greater than)
== ~=	(boolean equality/inequality)
& |	(boolean and/or)

Intrinsic functions

The MAST language provides the mathematical functions described below.

Table A-2. Intrinsic mathematical functions.

Function syntax	Description	Limitations
sin(x)	sine (x)	none
cos(x)	cosine (x)	none
tan(x)	tangent (x)	x cannot equal n(π/2)
asin(x)	arcsine (x)	$-1 \leq x \leq 1$
acos(x)	arccosine (x)	$-1 \leq x \leq 1$
atan(x)	arctangent (x)	returns a value between $\pm\pi/2$
sinh(x)	$\dfrac{e^x - e^{-x}}{2}$	none
cosh(x)	$\dfrac{e^x + e^{-x}}{2}$	none

Table A-2. Intrinsic mathematical functions. (continued)

Function syntax	Description	Limitations
tanh(x)	$\dfrac{(\dfrac{e^x - e^{-x}}{2})}{(\dfrac{e^x + e^{-x}}{2})}$	none
asinh(x)	$ln(x + \sqrt{x^2 + 1})$	none
acosh(x)	$ln(x + \sqrt{x^2 - 1})$	$x \geq 1$
atanh(x)	$ln(\dfrac{1+x}{1-x})$	
ln(x)	natural (base e) logarithm of x	$x > 0$
log(x)	common (base 10) logarithm of x	$x > 0$
exp(x)	e^x	$x \leq 80$
limexp(x)	numerically limited value of e^x	$x > 80$
sqrt(x)	square root of x	$x \geq 0$
abs(x)	absolute value of x	none
random()	generates a pseudo-random value between 0 and 1	takes no argument
d_by_dt(x)	differentiation with respect to time	can only be used in the equations section of a template; cannot contain **delay** or another **d_by_dt** operator

Table A-2. Intrinsic mathematical functions. (continued)

Function syntax	Description	Limitations
delay(x)	delay operator (e^{-sT})	can only be used in the equations section of a template; cannot contain **d_by_dt** or another **delay** operator

Declaration operators

Declaration operators provide a shorthand way to use composite parameters (such as structures), so they can be referenced without the need to write them out in full.

..   *Argdef* (argument definition)—allows "copying" of the declaration of an argument or parameter, usually from another template. For example, assume there is a structure named **model** containing several dozen number parameters in a template named **bjt**. To declare another parameter named **mymodel** that is identical to **model**, do the following:

         bjt..model mymodel

     The **mymodel** parameter is now declared as the same type of parameter as **model** (a structure), with all its subordinate parameters and their default values.

->   *Structure reference*—allows reference to a subordinate parameter within a union or a structure. For example, to refer to the parameter **rb** within

mymodel, do the following:

mymodel->rb

<-   *Structure overlay*—allows the definition of a new structure parameter by accessing individual parameters within another existing structure and giving them different values.
For example, to change the values of **is** and **rb** within **mymodel**, leave the other parameter values unchanged, and call the new structure **newmodel**:

newmodel=mymodel<-(is=4.7e-14, rb=9)

Assignment statements

An assignment statement uses an equals sign (=) to assign the value of a parameter, variable, or expression on the right hand side (RHS) to a variable on the left hand side (LHS). An assignment statement can appear only in the parameters section, the values section, or in a **when** statement, according to the following:

parameters section   In this section, only a local parameter can appear on the LHS of an assignment statement. The RHS usually consists of an expression containing a template argument. For example:

response = slewrate*1u
where **response** is a local parameter and **slewrate** is an argument.

values section   In this section, only a **val** can appear on the LHS of an assignment statement. The RHS usually consists of an expression containing any combination of variables, *except through variables*. For example:

vout = v(p) - v(m)

Reference Information **249**

	where **vout** is a **val**; **v(p)** and **v(m)** are across variables.
when statement	Within a **when** statement, only a local state variable can appear on the LHS of an assignment statement. The RHS usually consists of a scheduling assignment for any combination of variables, *except through variables*. For example:

```
when(event_on(notify)) {
    next_low=schedule_event(time+
    hightime, out, l4_0)
}
```

If constructs (condition checking)

You can use an if construct to check conditions and use operators to evaluate expressions based on those conditions. An if statement usually appears on multiple lines and can contain **else** and **else if** "blocks" to define different conditional branches to be checked. If statements can be nested. There are two basic forms of an if statement—an if statement and an if expression.

if statement	The basic if statement adheres to more formal programming conventions, where the condition is separated from expressions by opening and closing braces, { }.
	
	For example:
	
	if (cnom==undef) {
	    cap=0.0
	}
	else if (cnom==inf) {
	    cap=1.0
	    }
	else cap=cnom

if expression   This is a compressed form of an if statement that uses a shorthand method to check conditions and evaluate expressions. For example, the following if expression is equivalent to the if statement given above:

**cap =if cnom == undef then 0.0 \
    else if cnom == inf then 1.0 \
    else cnom**

Simulator variables

A simulator variable (simvar) is a predefined variable usually used to detect information from a simulation and provide it to the template. The main use of a simvar is to return a value of 0 or 1, which means it can be used in the condition portion of an if statement or a **when** statement to evaluate expressions based on simulator activity.

In addition, there are two simvars (**next_time, step_size**) that provide information from a template to the simulation. These simvars are used to control the occurrence and size of simulation time steps from within a template. The simulator variables used between MAST and Saber are:

dc_domain	freq	next_time	time_domain
dc_done	freq_domain	statistical	time_init
dc_init	freq_mag	step_size	time_step_done
dc_start	freq_phase	time	tr_done
			tr_start

Foreign functions

You can include a call to a foreign subroutine that performs calculations externally and returns them to the template. Note that the MAST language does not call a foreign routine directly—the call is interpreted by the simulator. Consequently, a special interface from the simulator to the foreign routine is required. In particular, the Saber simulator supports foreign subroutines written in C and in Fortran.

A foreign subroutine can return either a single number (which can be assigned to a template variable) or an unrestricted list of variables. In either case, the name of the subroutine must first be declared as a local variable. The returned value or values can then be assigned in the values or parameters section.

For example, the following excerpt shows how a to declare and use a subroutine that returns a single number. The name of the routine is **degtorad**; it converts degrees of phase to radians.

```
number    amp=1,           # argument for amplitude in volts
          freq=undef,      # argument for frequency in hertz
          phase=90         # argument for phase in degrees
{
foreign number degtorad ()   # declare subroutine
val v voltage                # declare calculated voltage
values {
voltage = amp*sin(2*pi*freq + degtorad(phase)) # use subroutine
...
}
```

MAST functions

It is sometimes advantageous to implement a particularly complicated portion of a template as a separate function that resides in its own file. Although similar to using a foreign subroutine, an external MAST function has the following advantages over using a foreign routine:

- The interface between function and template is substantially easier to write and is less error-prone.
- A MAST function can be debugged more easily.
- Porting to different computers is eliminated—the code is written in the MAST language, which is executed on each machine that the simulator is available on.
- The simulator interprets a MAST function more readily than it does a foreign function, which generally improves simulation efficiency.

Although a MAST function can perform a variety of tasks, it adheres to some of the same general format and syntax requirements of a template. It contains a header, header declarations section, and body; it resides in a file with the .sin extension and should be placed in the data search path. The MOSFET example given in Chapter 9 illustrates how a template uses a MAST function; the actual listings for those functions are given in Appendix B.

A.2 Newton Steps

As described in Chapter 1, a simulator finds the solution of nonlinear networks by solving a set of nonlinear, simultaneous equations using the Newton-Raphson iterative algorithm. While this approach has been found to be fairly robust for circuit simulation, it does require assistance for simulating circuits that employ sophisticated nonlinear device models. One of these forms of assistance is the newton step. Newton steps are not a unique concept to MAST and the Saber simulator. SPICE-based simulators [1] utilize newton steps in their nonlinear device models. Thus, newtons steps are not related to the *linearization techniques* employed by the simulator.

Newton steps tend to be empirically derived, but can have a dramatic effect on the convergence properties of a nonlinear model. The purpose of newton steps is to place a limit on the change of the independent variable from one iteration to the next. The effect of this is to restrict the range of approximation the simulator performs around the crossover points, which helps improve simulation efficiency.

Newton steps were first introduced in Chapter 8 in the **vlim** template. The voltage limiter model developed there used the MAST construct for implementing them. This example will be further described here.

Figure 8-2 shows that the dependence of **vout** on **vin** is piecewise linear, with -**vmax** and **vmax** defining crossover points for three separate regions of linear operation. For templates with this kind of input/output relationship, it is recommended that *newton steps* be specified for the independent variable (**vin**). When the independent variable is in a flat region, newton steps prevent the simulator from "guessing" a solution that grossly overshoots the actual solution. Such overshoots can cause slow convergence to a nonlinear solution or even numerical oscillation. Newton step increments are chosen

Reference Information

to be large enough to let the independent variable move from one piecewise linear segment to another, but small enough to prevent it from moving too far and possibly skipping a segment altogether.

Newton steps are related to the *iterative algorithm* that the simulator uses to find the solution of nonlinear equations. If these equations include exponentials, convergence may be slow, because a small change in the independent variable of the exponential may cause a large change in the function value.

More specifically, the goal is for the value of the independent variable **vin** to move quickly into the intended region of operation and, once there, have its movement restricted so that it is unlikely to leave the region again.

Specifying newton steps

Newton steps require three different statements to be included in the template:

1. A declaration of a structure parameter (**nvin**) to specify values for breakpoints and increments, as pairs of numbers in an array (**bp, inc**). This parameter may be declared either as an argument in the header declarations or as a local parameter in the template body. Values for these pairs are specified as described in Step 2, below.

   struc {
       number bp, inc
       } nvin[*]

2. An assignment statement in the parameters section that specifies values for **nvin**:

   nvin = [(-vmx,1.9*vmx),(vmx,0)]

3. A newton_step statement in the control section to associate the newton steps parameter (**nvin**) with the independent variable of the template (**vin**) :

   control_section {
       newton_step(vin,nvin)
       }

The meaning of the (breakpoint, increment) pairs is best defined by explaining the two pairs given for **nvin** in the assignment statement:

[(-vmx,1.9*vmx),(vmx,0)]

- Below the first breakpoint (**-vmx**), there is no restriction on how much **vin** can change from one iteration to the next.
- Between the first two consecutive breakpoints (**-vmx** and **vmx**), the change in **vin** is restricted to the first specified increment (**1.9*vmx**) per iteration.
- Above the last breakpoint (**vmx**), there is no restriction on how much **vin** can change.

To see why newton steps are used for this type of model, refer again to Figure 8-2. Typically, the solution of the nonlinear equations should be in the nonlimiting (central) linear region. If, during iterations, there is limiting (say, on the left side), it is not desirable to have **vin** "step over" the nonlimiting region and go directly to the limiting region on the right side. Instead, it is preferable to limit changes in **vin** such that it is in the nonlimiting region for at least one iteration.

Newton steps that have breakpoints (such as **-vmx** and **vmx**) that depend on the value given to an argument (**vmax**) are referred to as *parameterized*.

To accomplish this, newton steps are specified as shown for the nonlimiting region (between -**vmx** and **vmx**) but not for the upper and lower limiting regions. This has the effect of limiting the distance the simulator can step between ±**vmx** (i.e., when it enters the nonlimiting region or is inside the region) to 1.9•(**vmx**). Because this region has width 2•(**vmx**), *this newton step array prevents the simulator from stepping completely over the nonlimiting region.*

In general, the maximum allowable change should be less than the width of the critical region. In this example, there is no restriction on the size of an iteration step if **vin** remains in either the upper or lower limiting region. This is indicated by the fact that **nvin** does not specify limiting below -**vmx** or above +**vmx**. (The 0 increment means no limiting above +**vmax**.)

Example

Assume that **vmax** has been specified by the user as 10V, which sets the lower limit of the output to -10V and the upper limit to +10V, as shown in Figure A-1. Further, assume that **vin** is in the lower limiting region at -35V and that the iterative algorithm intends to change it to +35V. This would result in the simulator stepping over the nonlimiting region between ±10V.

Inside the nonlimiting region, the amount of change is restricted to 1.9•**vmax**, which is 19V. However, this restriction does not affect the amount of change *outside* the nonlimiting region (i.e., **vin** can move from -35V to +9V in one iteration; the 19V limit does not apply until **vin** crosses -10V).

Therefore, for the next iteration, **vin** will have a value of +9V (-10 + 19), which is in the nonlimiting region. Assuming that, based on the results of the iteration at +9V, the Newton-Raphson algorithm still intends to proceed to +35V, the value of **vin** would be +35V. This is because **vin** is not limited once it proceeds beyond +10V.

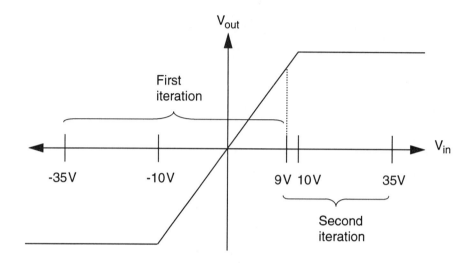

Figure A-1. How newton steps limit the change of **vin**.

A.3 References

1. L. W. Nagel and D. O. Pederson, "Simulation program with integrated circuit emphasis (SPICE)," *ERL Memo. No. ERL-M520*, University of California, Berkeley, May 1975.

APPENDIX B

MAST Template and Function Listings

This appendix provides the MAST listings for templates and functions of selected models described in this text. They have all been tested and, barring typographical errors, perform as intended with the Saber simulator from Analogy, Inc.

Chapter 4—Behavioral deadband

```
template deadband in out = thr, k
    ref nu    in              # Input (must refer to a "nu" var in another template)
    var nu    out             # Output
    number    thr = 1,        # Threshold value
              k = 1           # Transfer function gain (slope)
{
val nu outval
 struc sa {
      number bp, inc
      } ns[*]                 # Newton step array for "in"
 values {
     if (in >= -thr & in <= thr)    outval = 0
     else if (in > thr)             outval = k * (in - thr)
```

257

```
        else if (in < -thr)            outval = k * (in + thr)
        }
    control_section {
        newton_step(in,ns)
        }
    equations {
        out :  out = outval
        }
}
```

Chapter 4—Netlist of final design (Figure 4-8a)

```
v.input 0 cmd_volts = tran=(pulse=(v1=0,v2=10,tr=0.1,tf=0.1,td=1,pw=15,per=20))
rtnl2var.1 ang2 0 mid
var2elec.1 mid ang2_volts 0 = k=10
frictn_r.dchold ang2 0 = fric=1u, stic=1.1u
damper_r.shaft1d ang1 0 = d=10m
damper_r.shaft2d ang2 0 = d=10m
moi_r.shaft1j ang1 = j=5m
moi_r.shaft2j ang2 = j=1m
blsh_eff.mech ang1 ang2
r.div 0 sump = rnom=1k
r.zdiv2 0 mmid = rnom=1k
r.zdiv1 pmid 0 = rnom=1k
r.angfb ang2_volts summ = rnom=1k
r.in sump cmd_volts = rnom=1k
r.gainin gainm db_out = rnom=10k
r.gainfb gain_out gainm = rnom=3k
r.rtau mot_drive gain_out = rnom=10k
r.fbsum sum_out summ = rnom=1k
dc_pm.dcmot1 torq_volts 0 vel1 = ke=1, kt=1, laa=10m, j=1m, ra=1, d=1m
conv_r.w2a vel1 0 ang1 0
vcvs.mdrive 0 torq_volts 0 mot_drive = k=-1
c.ctau 0 mot_drive = c=10u
oab.sum sum_out summ sump 0 = a=1meg, vos=0, rout=10, rin=100k, fu=100k, f2=1meg
oab.gain gain_out gainm 0 0 = a=1meg, vos=0, rout=10, rin=100k, fu=100k, f2=1meg
d.dpos db_out pmid
d.dneg mmid db_out
zd.zneg mmid sum_out = model=(vzt=0.5,izt=1e-3,rzt=1)
zd.zpos sum_out pmid = model=(vzt=0.5,izt=1e-3,rzt=1)
```

Chapter 6—Behavioral pole-zero transfer function

```
template pz vip vim vop vom = a, fp, fz, init
electrical vip,vim,vop,vom
number       a  = undef,
             fp = undef,
             fz = undef,
             init = undef
{

    var i    i
    val v    vin,vout,
             voutb,voutp,vinc,vinz,outdc

#...outdc...output value held during a dc analysis if used as an integrator
    number tz=0.0,tp=0.0,c=1.0,b=0.0,d=1.0

#****************************************************************
# Implemented transfer function (in terms of the above numbers) is:
#                    (c + tz*s)
#   H(s) = vout/vin = ---------------------
#                    b*(d + tp*s)
#****************************************************************

parameters {
    if (a == 0.0d0) {
            c = 0.0
            b = 1.0
            }
    if ((a ~= undef) & (a ~= 0.0d0)) {
            b = 1/a
            }
    }

values {
    vin   = v(vip) - v(vim)
    vout  = v(vop) - v(vom)
    voutb = d*b*vout
    voutp = tp*b*vout
    vinc  = c*vin
    vinz  = tz*vin
    outdc = 0
```

```
        #...integrator output at dc
        if (dc_domain & init ~= undef & fp == 0.0d0) {
                outdc = vout
                vinc  = init
                }
        else {
                outdc = 0
                }

        }

equations {
    i(vop) += i
    i(vom) -= i
    i: outdc + voutb + d_by_dt(voutp) = vinc + d_by_dt(vinz)
    }
}
```

Chapter 9—MOSFET MAST function mosparam

The following MAST function performs parameter calculations and validity checking for the template **mos**.

```
function work = mosparam(model,temp,input_l,input_w)
mostype..model      model
mostype..work       work
number              temp, input_l, input_w
{
<consts.sin
number  xkt,fc,
        temp2,tnom,                     # for temp ~= tnom
        trat,trat3,trat5,               # for temp ~= tnom
        kt2,vtnom,                      # for temp ~= tnom
        egt2,egtnom,                    # for temp ~= tnom
        vjfact,phinom,efact,vjnom       # for temp ~= tnom
number  eps0 = 8.854214871d-12,
        epsox
epsox = 3.9*eps0
# Transfer the values in the parameter list to the work list if
# they are modified in the program.
work->wvto = model->vto
work->wkp = model->kp
```

```
work->wgamma = model->gamma
work->wphi = model->phi
work->wlambda = model->lambda
work->wis = model->is
work->wvj = model->pb
work->wcj = model->cj
work->wfc = model->fc
# Calculate the temperature in Kelvin.
tnom = model->tnom + math_ctok
# Calculate some values.
xkt = math_boltz*tnom
work->wvt = xkt/math_charge
if(model->ld ~= undef) work->wleff = input_l - 2*model->ld
if(model->wd ~= undef) work->wweff = input_w - 2*model->wd
# Set the type base on the TYPE flag.
work->wtype = 1
if (model->type == _p) work->wtype = -1
if (model->tox ~= undef & model->tox ~= 0) work->wcox = epsox/model->tox
# CGBO
if ((model->cgbo ~= undef) & (model->cgbo >= 0)) {
    work->wcgbe = model->cgbo*work->wleff
    }
else if ((model->wd ~= undef) & (model->wd >= 0)) {
    work->wcgbe = work->wcox*model->wd*work->wleff
    }
else {
    work->wcgbe = 0
    }
# CGDO
if ((model->cgdo ~= undef) & (model->cgdo >= 0)) {
    work->wcgde = model->cgdo*(work->wweff + 2*model->wd)
    }
else if (model->ld ~= undef & model->wd ~= undef) {
    work->wcgde = work->wcox*model->ld*(work->wweff + 2*model->wd)
    }
else {
    work->wcgde = 0
    }
# CGSO
if ((model->cgso ~= undef) & (model->cgso >= 0)) {
    work->wcgse = model->cgso*(work->wweff + 2*model->wd)
```

```
        }
else if (model->ld ~= undef & model->wd ~= undef) {
    work->wcgse = work->wcox*model->ld*(work->wweff + 2*model->wd)
        }
else {
    work->wcgse = 0
        }
work->wfunc2 = work->wvto-work->wtype*work->wgamma*sqrt(work->wphi)
if (model->tnom ~= temp) {
    # Set the temperature terms
    temp2 = temp + math_ctok
    trat = temp2/tnom
    trat3 = trat**3
    trat5 = sqrt(trat3)
    kt2 = math_boltz*temp2
    vtnom = work->wvt
    work->wvt = kt2/math_charge
    egt2 = 1.16D0-(7.02D-4*temp2*temp2)/(temp2+1108)
    egtnom = 1.16D0-(7.02D-4*tnom*tnom)/(tnom+1108)
    work->wkp = work->wkp/trat5
    vjfact = -3*work->wvt*ln(trat) + egt2 - egtnom*trat
    phinom = work->wphi
    work->wphi = work->wphi*trat + vjfact
    work->wfunc2 = work->wfunc2 + work->wtype*0.5*(work->wphi - phinom) -
            0.5*(egt2 - egtnom)
    work->wvto = work->wfunc2 + work->wtype*work->wgamma*sqrt(work->wphi)
    efact = trat3*limexp(-egt2/work->wvt + egtnom/vtnom)
    work->wis = work->wis*efact
        }
work->wbeta = work->wkp*work->wweff/work->wleff
work->wcox = work->wcox*work->wweff*work->wleff
work->wfunc2 = work->wtype*work->wfunc2
}
```

Chapter 9—MOSFET MAST function mosval

The following MAST function computes the drain-source current and the channel charges for the template **mos**.

```
function (ids,qg,qd,qs,qb) = mosval(work,model,vgs,vds,vbs)
val i         ids
```

MAST Template and Function Listings

```
val q         qg,qd,qs,qb
mostype..model      model
mostype..work       work
val v         vgs,vds,vbs
{
val v         vds_arg,vbs_arg,vgs_arg,vfb,vgd,vbd,vgb
val v         von,vgst,vdsat,vthresh,vgdt,vgbt,vgsdt
val nu        betap,sarg
if(vds>=0) {
    #...Normal mode
    vds_arg=vds
    vbs_arg=vbs
    vgs_arg=vgs
    }
else {
    #...Inverse mode
    vds_arg=-vds
    vbs_arg=vbs-vds #...vbd
    vgs_arg=vgs-vds #...vgd
    }
#...Call the appropriate routine
vbd = vbs-vds
vgb = vgs-vbs
vgd = vgs - vds
vfb = work->wfunc2 - work->wphi
if (vbs <= 0) {
    sarg = sqrt(work->wphi-vbs)
    }
else {
    sarg = sqrt(work->wphi)
    sarg = sarg - 0.5*vbs/sarg
    sarg = max(0,sarg)
    }
von = work->wfunc2 + work->wgamma*sarg
vgst = vgs - von
vdsat = max(vgst,0)
# Calculation of drain current A. cutoff region
if (vgst <= 0) {
    ids = 0
    }
else {
```

```
        betap = work->wbeta*(1+ work->wlambda*vds)
        if (vgst > vds) {      # B. linear region
                ids = betap*vds*(vgst - 0.5*vds)
                }
        else {                 # C. saturation region
                ids = 0.5*betap*vgst**2
                }
        }
#...Calculate amended threshold voltage
if(vbs <= 0) {
        vthresh = von
        }
else if(vbs <= 2*work->wphi) {
        vthresh = vfb + work->wphi + work->wgamma*(sqrt(work->wphi) -
                vbs/(2*sqrt(work->wphi)))
        }
else {
        vthresh = vfb + work->wphi
        }
vgst = vgs - vthresh
vgdt = vgd - vthresh
vgbt = vgb - vthresh
#...Below Flat Band region
if(vgb <= vfb) {
        qg = work->wcox*(vgb - vfb)
        qb = -qg
        qd = 0
        qs = 0
        }
#...Below Threshold region
else if( (vgdt <= vgst) & (vgst <= 0) ) {
        qg = work->wcox*work->wgamma*(sqrt((work->wgamma*work->wgamma/4) +
                vgb - vfb) - work->wgamma/2)
        qb = -qg
        qd = 0
        qs = 0
        }
#...Saturation region
else if( (vgdt <= 0) & (0< vgst) ) {
        qg = work->wcox*((2/3)*vgst + (von - work->wfunc2))
        qb = -work->wcox*(von - work->wfunc2)
```

```
        qd = -(4/15)*work->wcox*vgst
        qs = -(2/5)*work->wcox*vgst
        }
#...Triode region
else if( (0 < vgdt) & (vgdt <= vgst) ) {
        vgsdt = vgst + vgdt
        qg = work->wcox*(von - work->wfunc2 + (2/3)*(vgsdt -
                vgst*vgdt/vgsdt))
        qb = -work->wcox*(von - work->wfunc2)
        qd = -(work->wcox/3)*(vgdt/5+ 4*vgst/5 +
                vgdt*vgdt/vgsdt + vgst*vgdt*(vgdt - vgst)/(5*vgsdt*vgsdt) )
        qs = -(work->wcox/3)*(vgst/5+ 4*vgdt/5 +
                vgst*vgst/vgsdt + vgst*vgdt*(vgst - vgdt)/(5*vgsdt*vgsdt) )
        }
# Take into account inverse mode of operation
if(vds<0) {
        ids = -ids
        }
# Take into account p or n type.
ids = work->wtype*ids
qg = work->wtype*qg
qd = work->wtype*qd
qs = work->wtype*qs
qb = work->wtype*qb
}
```

Chapter 9—MOSFET MAST function mosdiode

The following MAST function computes the current and charge of the bulk diodes for the template **mos**.

```
function (id,qd) = mosdiode(work,model,vd)
val i           id
val q           qd
mostype..work           work
mostype..model          model
val v           vd
{
number   evd,gd,arg1,sarg,arg2,arg
# Find the current from the bulk to source/drain diode.
if (vd > 0) {
        evd = limexp(vd/work->wvt)
```

```
        gd = work->wis*evd/work->wvt+model->gmin
        id = work->wis*(evd-1) + vd*model->gmin
        }
else {
        gd = work->wis/work->wvt
        id = gd*vd + vd*model->gmin
        }
if (vd >= work->wfc) {
        arg1 = 1 - work->wfc/work->wvj
        sarg = limexp(-model->mj*ln(arg1))
        arg2 = work->wcj*sarg*model->mj
        qd = work->wvj*work->wcj*(1-arg1*sarg)/(1-model->mj) +
                (vd-work->wfc)*(vd-work->wfc)*arg2/(2*work->wvj*arg1) +
                (vd-work->wfc)*(work->wcj*sarg)
        }
else {
        arg = 1 - vd/work->wvj
        sarg = limexp(-model->mj*ln(arg))
        qd = work->wcj*(1 - arg*sarg)/(1 - model->mj)
        qd = work->wvj * qd
        }
id = work->wtype*id
qd = work->wtype*qd
}
```

Index

A

absolute value 91
abstraction, level of 39, 67
across variable 4, 15, 55, 64, 79
   hydraulic 200
   magnetic 223
   mechanical 174, 175
   thermal 207
AHDL 4, 13, 67
analog state 135, 157
analog-to-digital converter 144
angle, as across variable 175
argdef 247
argument 52, 59, 62, 242
   passing through hierarchy 72
array 94, 244
assignment statement 248

B

backlash, mechanical 43
behavioral models 19, 20, 34, 43, 45, 87, 193
B-H curve 229
binary operators 245
body 61
Boyle op amp model 7, 20, 29

brace 239
branch 65, 79, 84
bulletproofing 62, 81

C

characteristic equation 58, 64
characterization 10
charge 82
   conservation of 15
   diode 99
charge-control models 21
circuit build-up macromodel 28
coefficients 16, 62
collapse statement 105
comma 85
comment 59
component model 10
compressibility 200
conceptual model 8
condition checking 80, 249
conditional statement 92
conductance 94
conflict resolution 130
   at digital nodes 131
   table for logic_4 132
connection points 59, 145, 176, 240

digital 124
   input 147
   output 147
conservation of a through
      variable 4
conservative model 65
constant.sin file 241
constraint equation 66
continuous time simulation 133
control section 63, 155, 254
   diode 105
   MOSFET 117
control system 146
core 225
correlation 164
cumulative density function
      (CDF) 163
current source, constant 77
current, as through variable 55,
      58, 79

D

d_by_dt operator 83, 151, 156
   using with multiplication 151
data flow 146
data table, one-dimensional 88
data type 52, 243
data-based model 87
DC domain 13
DC initialization, of a state
      variable 141
dc_init statement 137
dc_init variable 126
deadband 44
declaration 52
   header 61
default value 78, 91, 125, 141
   not provided 80

delay 135
dependent variable 4, 15, 55, 58
differential equations 4, 15, 16,
      64
   behavioral implementation 34
   simultaneous 32
differentiation 82, 83, 151
digital 14
   initialization 125
   modeling 121
   signal 122
   time 122
digital HDL 13
digital-to-analog converter 144
diode 98
   power 21
   zener 22
discontinuity 55, 92
discretized partial differential
      equation-based models 21
driven function 131
dynamic thermal model 212

E

efficiency, simulation 31, 69,
      79, 119
electromagnetic devices 224
electro-mechanical models 173
   motor 186
electro-thermal model 212
electro-thermal simulation 205,
      206
emf, back 191
enum 243
equals sign (=) 59
equations section 64
   diode 104
   MOSFET 117

Index

equations, setting up 16
event 14
event queue 63, 122
event_on function 125, 141
event-driven 14, 16, 121, 133, 240
event-driven analog 135
expression 244

F

finite element 21
   mechanical 173
floppy disk drive system 187
flow, as through variable 200
fluid power system 202
flux density 228
flux, as through variable 223, 227
force, as through variable 175
foreign routine 88, 95, 251
friction 43
   rotational 192
   static 192
function
   built-in 88
   MAST 110, 251
functional block 145
functional models 19

G

Gaussian elimination 29
Gear technique 6
gear, mechanical 179
gearbox, mechanical 41
generator 185
generic model 9
glue chip 11

H

header 59
heat sink 207, 212, 213, 217, 220
hierarchical template 73
hierarchy 68
hunting, mechanical 43
hydraulic models 199
hysteresis, magnetic 230, 235

I

if construct 80, 249
if expression 250
if statement 249
IGBT 205
   thermal network for 216
implicit equations 15
include file 103, 241
independent variable 4, 15, 55, 57, 93
initial condition 15, 105, 116
initial_condition statement 155
initialization
   DC 155
   digital 127, 129, 138
input file 51
input, connection point 147
instance 53
integration 85, 154
intrinsic functions 245

J

junction temperature 211

K

KCL 4, 15, 65, 79, 147, 174, 207
Kirchhoff's current law 4, 15
Kirchhoff's voltage law 15
KVL 15, 65

L

Laplace transform 39, 151
large-signal 16, 97, 104, 108
large-signal model
  diode 99
linear devices 77
linear model 88
linearization techniques 252
line-continuation characters 240
local declarations 61
  diode 103
  MOSFET 116
local parameter 81, 243, 248
logic levels 122
logic_4 122, 242
  conflict resolution table 132
loss-of-step 193
low-pass filter 70

M

macromodeling 20, 27, 34
magnetic field strength 229
magnetic models 223
magnetomotive force (mmf) 223
make 66, 85, 153
MAST function 251
mathematical functions 245
matrix 4, 54

mean, statistical 161
mechanical domain 40
mechanical modeling 174
mechatronics 173
mixed analog-digital ASIC 188
mixed analog-digital simulation 133
mixed-signal simulation 144
mmf 226
mmf, as across variable 223
.MODEL 103
model
  conceptualization 51
  conceptualizing 16
  development 56
  implementation process 17
  philosophy of writing 55
  validation 23
modeling hierarchy 18
modified nodal analysis 55
modularity 68
Monte Carlo analysis 160
MOSFET 10, 15, 106, 182
  operational characteristics 118
MOSFET, power 11
motor 179, 185, 216

N

netlist 4, 52, 59, 68, 70
  usage 57
netlist entry 72
netlist section 62
newton step 93, 119, 252
  parameterized 254
newton_step statement 93, 105, 254
Newton-Raphson 6, 252

Index

node 52, 124
   internal 103, 116
noise analysis 165, 168
nonlinear magnetics 229
nonlinear model 89, 97
normal distribution 160
number 243
numerical integration 6

O

Ohm's law 57, 79
output, connection point 147

P

parameter 52, 58, 242
   temperature dependence 62
parameters section 62, 248
   diode 104
   MOSFET 116
parasitics, mechanical 41
partial derivatives, of semiconductor model 208
permeability 228
piecewise linear interpolation 88
pin definitions 242
pins 240
position, as across variable 175
post-simulation statements 64
pound sign (#) 59, 240
power diode 207
power dissipation 212
power supply circuit 232
power, as through variable 207
pressure, as across variable 200
primitive model 19, 29, 39

probability density function (PDF) 160
probability distribution 160
procedural section 63
prototyping 38
pseudo-template 171, 177, 186, 200, 210, 230
   hose 201
   igbt_th 211
   rack_pin 178
   stp_2ppm 187
   wndg 231
PWM voltage-source inverter 214

R

random number function 165
ref 145
ref connections 147
reference designator 53, 72
reserved word 52, 85
right hand rule 228

S

Saber simulator 6, 14, 69, 132, 165, 174, 209, 241, 252
   DC algorithm 137
Safe Operating Area (SOA) 8
sampled data systems 157
schedule_event function 126
schedule_next_time statement 142
schematic entry program 72
s-domain 150
seat position motor drive circuit 179
section, template 52, 59

self-heating 206, 209
semicolon 239
semiconductor device model 97
shaft angle controller 39
shaft, flexible 43
signal flow 146
simplification macromodel 28
simulation factors, in a model 58
simulation process 55
simulation, continuous time 15, 54
simulator variable 126, 127, 250
simvar 250
small-signal parameters 105
SPICE 6, 14, 68, 106, 174, 205, 252
  netlist 103
standard deviation 161
standardization 25
state 122, 124
  analog 135, 141
  variable 135
static thermal model 212
statistical modeling 159
stepper motor 187, 190
stiction 192
string 243
struc 243
structural representation 35
structure 243
structure overlay 248
structure reference 248
subcircuit 68
.SUBCKT 68
switch, digitally controlled 140
switching losses 217
symbolic macromodel 29

syntax 72, 239
system variable 15, 35, 121, 122, 124

T

table-based model 87
technologies, non-electrical 18
temperature 58, 62, 103, 182
  external 167
temperature, as across variable 207, 208
template 9, 51, 52, 59, 68
  and 130
  body 61
  capacitor 82
  clock 128
  comparator 134
  dcsrc 148
  deriv 152
  diode 10, 101
  dlay 157
  header 59
  hierarchical 72
  inductor 86
  inverter 123
  isource 78
  low_pass 73
  mos 112
  mostype 115
  pwlcond 94
  resistor 80, 81
  resistor_ns 169
  section 59
  sections 103
  sum2 149
  sumz 159
  sw 140
  vlim 90

Index

vsource 85, 148
template sections 64
terminal characteristics of model 64
thermal coupling 206
thermal models 205, 208
thought process, modeling 51, 97
threshold function 126, 136, 142
through variable 4, 15, 55, 64, 79, 249
  hydraulic 200
  magnetic 223
  mechanical 175
  thermal 207
time 250
  delay 156
  digital 122, 126
time step 92, 136, 142
time_init variable 126
tolerance value 161, 164
top-down design 37, 70, 145
topology 16, 29, 99, 103, 104, 108
torque, as through variable 175
tr_start variable 126
transfer function 39, 40
  two-pole 155
trapezoidal technique 6

U

unary operators 245
undef 155
uniform distribution 160, 164
union 244
unit definitions 241
units 83

magnetic 226
mechanical 176
thermal 210
units.sin file 241

V

val 61, 63, 82, 83, 92, 166, 169
validation 24
values section 63, 249
  diode 104
  MOSFET 117
var 61, 145, 147, 169
  current through a noise voltage source 168
voltage source, constant 84
voltage, as across variable 55, 57, 79

W

when statement 63, 125, 135, 249
  setting values within 141
  used in a netlist 143
winding 225

Z

z-domain 157
zener diode 22
  bridge 46
z-transform 158